Style and Time

AGM STUDIES

AVANT-GARDE & MODERNISM

Style and Time

Essays on the Politics of Appearance

ANDREW BENJAMIN

Northwestern

University Press

Evanston

Illinois

Northwestern University Press
www.nupress.northwestern.edu

Printed in the United States of America

10 9 8 7 6 5 4 3 2 1

ISBN 0-8101-2333-9 (cloth)
ISBN 0-8101-2334-7 (paper)

Library of Congress Cataloging-in-Publication Data

Benjamin, Andrew E.
Style and time : essays on the politics of appearance / Andrew Benjamin.
p. cm. — (Avant-garde & modernism studies)
Includes bibliographical references and index.
ISBN 0-8101-2333-9 (cloth : alk. paper) — ISBN 0-8101-2334-7 (pbk. : alk. paper)
1. Benjamin, Walter, 1892–1940—Aesthetics. 2. Style (Philosophy) 3. Time. 4. Interruption (Psychology) I. Title. II. Series: Avant-garde and modernism studies.
B3209.B584B445 2006
111.85—dc22

2005023274

♾ The paper used in this publication meets the minimum requirements of the American National Standard for Information Sciences—Permanence of Paper for Printed Library Materials, ANSI Z39.48-1992.

Contents

Acknowledgments, vii

Introduction, ix

Part One. Working through Walter Benjamin

1 Benjamin's Modernity, 5

2 The Time of Fashion: A Commentary on Thesis XIV in Walter Benjamin's "On the Concept of History," 25

3 Benjamin's Style: The Style That Is Not Jugendstil, 39

Part Two. Unforeseen Appearances

4 The "Place" of Cosmopolitan Architecture, 63

5 In What Style Should We Build? The Style of Cosmopolitan Architecture, 80

6 Refugees, Cosmopolitanism, and the Place of Citizenship, 107

7 The Matter of a Materialist Philosophy of Art: Bataille's Manet, 124

Notes, 139

Index, 165

Acknowledgments

I would like to thank the publishers for permission to reprint the following parts of this book, earlier versions of which were previously published: chapter 1, from *The Cambridge Companion to Walter Benjamin,* edited by David Ferris (Cambridge University Press, 2004); chapter 2, from the journal *Thesis* 11 (no. 75, 2003); and chapters 4 and 6, from *Architectural Theory Review* 7 (nos. 1 and 2, respectively, 2002).

This book has been made possible with the assistance of an Australian Research Council Grant. It forms part of the project "Between the Outback and the Sea: The Place of Cosmopolitanism in Contemporary Australia," undertaken at the University of Tasmania.

Finally, I would like to thank Dimitris Vardoulakis for his help with the preparation of the manuscript.

Introduction

Writing on style, while attempting at the same time to distinguish between style and tradition, as well as style and fashion, may seem an undertaking whose relevance needs to be questioned. However, the contention to be advanced throughout these essays is that the question of style has never been more relevant. The basis of this claim is found in the intricate set of interconnections that define style; i.e., the complex of relations between appearance, recognition, and identification. The significance of each of these terms is that they indicate that issues having to do with style and presence cannot be thought outside their specific relation to forms of social identity arising from structures of recognition. Style, even at its most straightforward, involves a relationship between appearance and recognition. (Style has to be recognized as such.) The interconnections between these terms establish what will be called the *site* of style. Working through this site will open up what had been thought to have been closed down. Style will acquire a relevance that its conflation with either predetermined appearances on the one hand or fashion on the other would have rendered obscure. To deploy a formulation from the writings of Walter Benjamin, a passage that will continue to be of great significance throughout the discussion to come, what this will involve is an awakening of style. In *The Arcades Project* Benjamin writes that

> we construct an awakening theoretically—that is, we imitate, in the realm of language, the trick that is decisive physiologically in awakening, for awakening operates with cunning [*Erwachen operiert mit der List*]. Only with cunning, not without it, can we work free of the realm of dream. (*AP* 907/*GS* 5.2:1213)[1]

The language of dream and awakening is not a simple opposition. The decisive term that interrupts such a possibility is "cunning" *(List).* Cunning not only evokes freedom from the domain of dreams, it also enables the dream's actual nature to be recognized. While the term "cunning" has a history of great significance, it can be understood as introducing the realm of directed activity.[2] Awakening necessitates action, even if the form taken by that action awaits specification. The action in question involves a process, which in freeing itself from the hold of description opens other possibilities. While these other possibilities will have to be named, the important point to note at the beginning is that awakening involves, for Benjamin, a form of transformation. Awakening is linked as much to a remembering as it is to a type of projection. To awaken is to transform that which figures in the process. Both object and subject are transformed in the activity. In being remembered continuity is sundered, though in terms of another possibility. This possibility, however, should not be understood as a predetermined telos. It is an interruption that occasions. (Production does not occur *ex nihilo* but *ex interruptio.*) Part of what occasions is a form of interruption in which awakening is linked to truth. Not truth in an epistemological sense, but the truth about the nature of time.

The question of time warrants more attention than can be adequately given here. In sum, the assumption behind the arguments developed throughout this book—though particularly in the first three chapters—starts by rejecting the premise that historical time is inherently chronological and reciprocally that chronology is the temporality proper to the unfolding of history. The basis of this rejection is that holding to such a position would entail a type of immutability both in relation to events as well as their ordering. This position naturalizes a chronology sanctioning the repetition of the always the same. (Not the "always identical," but the same understood as that which can admit neither of difference nor interruption.) This positioning of time is in the service of one political position rather than another. If it is possible to argue for an interruption so that discontinuity marks the progress of time, then that possibility does not refer simply to the quality of events. It will inhere in the reality of historical time. Time must allow for its own interruption. A distinction can be drawn, therefore, between historical time—of which this is the most apposite description—and biological or geological time. Moreover, historical time cannot be reduced merely to the moments that mark the discontinuous presence of separate events. The possibility of overcoming both the temporality of separate events on the one hand, and a naturalization of continuity on the other, is found in the insistence of the discontinuous. What this insistence entails is that discontinuity must engender another account of continuity. Interruption is not out of time. (It follows, in addition, that its place is not utopian.)[3] The occurrence of discontinuity indicates that time allows for an interruption that yields another beginning; a beginning that cannot be accounted for in terms of the temporality of separate events. As such, time takes on the quality of the Absolute. The Absolute becomes the truth of time.[4] Understood in this sense, time allows for its own interruption, and the presence of interruption evidences that state of affairs. There cannot be a presentation of time as the Absolute, since this conception of time is the condition of possibility for presentations. Time as the Absolute is therefore individuated with particulars while allowing for overcomings and beginnings.

Taken overall, what this account of time means is that interruption—on one level a version of destruction—cannot be equated with the project of nihilism. The continuity in question, as well as transformation's capacity to engender a different account—the possibility of other narratives—has to be understood as claims that are as much about historical time as they are about the creation of those narratives. Both the object in its being awakened and the subject transfigured in the act have to bear the mark of that moment. The mark identifies and delineates; it allows for a recognition and identification.

While such an abstract formulation of time seems to be far removed from any concern with style, even style as a historical phenomenon, this is not the case. The site of style can be defined, as has already been suggested, as the relationship between appearance, recognition, and identification. This is, of course, a minimal description since each term—"appearance," "recognition," and "identification"—stands in need of clarification. What should be noted from the start, however, is that each one of these terms has a complex though necessary temporal dimension, since each one evidences conditions of historical time. In general terms, what this means is that the question of style is always interarticulated with questions of time and transformation; each moment intrudes into the other. This intrusion will be the case even if transformation is only ever present as an immanent possibility.

Style has a twofold designation. It locates particularity; the particular will have a certain style. At the same time, however, it sustains generality, since the particular may be the particular instance of the general. In broader terms, therefore, style becomes a way of marking the subservience of an object to a specific tradition. Precisely because of this relation, style can be understood equally as that which marks, or may mark, the refusal of a tradition. Style, thus construed, is in the service of any thinking of historical time. There is a further point of fundamental importance that emerges from the necessary connection of style and tradition: namely, that any concern with style is automatically a concern with the public realm. Style, even if it were taken to be idiosyncratic, is nonetheless and by definition connected to the domain of the public. The reason is that style appears. Style operates as an appearance precisely because of its connection to recognition. In raising the question of recognition, any appearance automatically brings the site of recognition into play. That site is the domain of the public. (It is the very presence of style as an appearance—and one occurring in the public realm—which, as will be argued, opens up its initial connection to the symbol.) Despite the recourse to a language of appearance and thus to a possible explanation on the level of appearance, questions endure. What is it that appears? What is appearance? What is style?[5]

Benjamin can be taken as pointing the way toward a consideration of these questions when, in relation to Jugendstil, he writes in *The Arcades Project:*

> One could try to show, using the example of Baudelaire, that Jugendstil arises out of weariness—a weariness that manifests itself, in this case, as that of the mime who has taken off his makeup. (*J* 82, 5)

Style, therefore, to the extent that this comment allows for a type of generalization, occurs as the result of a cultural mood. In addition, the mood in question is not one that is linked to an essentializing conception of culture. "Weariness" here introduces an understanding of culture defined by internal divisions and thus having the inevitability of conflict.[6] The very presence of such conflicts must complicate the question of style and specifically the relationship between style and appearance, since it undoes the unifying exigency that was taken to define style. An exigency that it effected through appearance. Part of the reformulation of style is that once its necessary connection to an already determined appearance is called into question, style will have become a realm in which the questions of what appears and how appearance is to be understood will remain as questions. The answers, in no longer being determined in advance, would signal the severance of both style and appearance, and then as a consequence, the repositioning of style as that which opens up appearance as a question. The question—what is style?—returns therefore at its most emphatic. (It should be noted, of course, that even at the beginning, given the reality of being able to separate questions of style from those of appearance, it is clear that the relation between style and appearance is not a necessary one, if necessity is understood as that which is determined in advance.)

In pursuing the question of style—pursuing it through the site of style—this book will address a number of interconnected moments. Despite apparent differences between them, there is a pervasive continuity of concern. In addition, the fact that the address allows for a type of discontinuity may itself reveal something about the nature of the topic. Perhaps the easiest way into the problematic of how style is to be understood—where the setting is the "present," the site of modernity itself—is to recognize that an integral part of the site of style is the complex relationship between modernity and symbols. The very unity that symbols demand—a unity that inheres in the conventions of style—is undone in advance by the inception of the modern. The symbol therefore becomes a vestige. And yet, since modernity cannot be separated from the struggle to affirm its own presence, these vestiges return within that contestation. The return, however, is of a vestige determined by its own impossibility. (It may be this very impossibility that accounts for the violence endemic within contemporary nationalism. Nationalism is the politics of the symbol par excellence, though now it is a politics structured by the inescapable failure of the symbol's unifying power.)[7] The difficulty of giving a definition of modernity is that the form modernity takes is not unitary. In sum, the modern emerges to the extent that the past is understood as irrecoverable. What is fundamental, however, is the nature of this irrecoverability. The past in question could be the distant past of antiquity, or it could be the relative past of an individual's life. What is essential is that this irrevocability be perceived. (And this will be the case even if that perception is either repressed or disavowed.) The question of modernity is, of course, more complex. Historicism—in Walter Benjamin's sense—becomes the attempt to establish continuity. This project of historicism occurs precisely because the interruptions that account for the modern are ever present. The attempt to preclude them on the one hand and the continual struggle to affirm them on the other are defining moments marking the advent of modernity.

This book consists of two parts, "Working through Walter Benjamin" and "Unforeseen Appearances." Taken overall, the book's project is to establish a series of interrelated positions. While they have already been intimated, they can be identified with greater precision. The first position is that the modern is bound up with an interruption or severance. Moreover, the force of tradition (evidenced in either a projected sense of national continuity or of historicism in general) means that the modern is a project that is always to be established. Hence, modernity names the site of contestation in which the project becomes the rearticulating of the modern. The second position is that a way into these concerns can be found by tracing the severance of style and appearance that marks the production of art and architecture in the modern period. The third and final position is that this severance is not singular in nature. It is as much present in the questions of stylistic continuity that define the central debate in German architectural theory in the nineteenth century as it is in the crisis of the symbol as the bearer of national unity, or as it is in the problem of precedence and hence continuity that occurs within painting with Manet. (Other instances could of course be cited.) The connection between these positions is in a sense quite straightforward; each one is evidence for the other. More is at stake, however, since the general argument is not an attempt to rekindle the quarrel of the ancients and the moderns. At work here is the interplay between historical time and action. Moreover, there are different and conflictual registers of historical time, and as a consequence any one way of conceiving of historical time (one way as opposed to another) will generate different actions, strategies, and policies. Indeed, the impossibility of assimilating conceptions of continuity and discontinuity means that their clash is further evidence of the hold of the modern. Affirming the discontinuous becomes the affirmation of modernity. On the other hand, however, arguments for continuity are not the denial of the modern, since they are occasioned by it. They need to be evaluated in terms of disavowal and thus the refusal of affirmation.

Working through Walter Benjamin

Walter Benjamin's project—or at least the elements that are central for this undertaking—inheres in the possibility that interruption and discontinuity can be understood as originating (or generative). In other words, it is a project that while privileging the caesura—the moment of interruption—does so on the basis that it is thought beyond the hold of nihilism. This does not make Benjamin's project Romantic—despite the centrality of the caesura for Hölderlin—what it indicates is that Romanticism, or at least currents within it, gave to modernity the temporality proper to it.[8] (Or at least this will be the argument.) The chapters constituting part I of this book chart a course through Benjamin's writings, a course privileging the interplay of time and interruption. This interplay involves a complex relation to memory understood as citation and repetition. The project is to work through Benjamin. Working through in this context involves two directions of activity: directions that continue to intersect both explicitly and implicitly. In both instances, time plays a fundamental role.

The first direction necessitates arguing that an important distinction can be drawn between the temporality of fashion—a temporality in which the new is simply posited—and the temporality of modernity. The latter, as has already been suggested, is marked by a sustained interruption, or rather an interruption that comes to be sustained by intellectual, cultural, and political action. Hence interruption brings with it an inherently productive potential. The second is that concomitant with this direction is another whose location has greater historical precision. In this instance, it is the severance between style and appearance: a position announcing the inception of the debate on style in German architectural theory in the nineteenth century (though, as will emerge, it is also the basis of Manet's painting). This severance means that the appearance of the modern is not determined in advance. It is not simply that imitation is no longer apposite; rather, it is the impossibility, within modernity, of there being ideals to imitate demands that art be linked to another sense of the appropriate. This sense will move propriety away from the content of artwork and toward the conditions of its reception. The presence of art has to be established. What this means is that what is proper to art is criticism. (Taking up Benjamin's engagement with Romantic criticism, art can be defined as that which is criticizable.) In more general terms, the overall significance of Benjamin's work on time is that it provides the basis for giving an adequate account of the severance of style and appearance and the consequences of this for artistic production as well as political practice. Benjamin's work is a prompt. Hence the description of "working through" Benjamin.

In order to develop an understanding of modernity defined in terms of historical time, a distinction needs to be drawn between a conception of the modern conceived as the current state of progress and modernity as an interruption. The first is the linear conception of development through continuity that is the implicit understanding of historical time in the Enlightenment, especially in the writings of Kant.[9] Working against this tradition involves deploying motifs from Romanticism, more specifically the conception of caesura in Hölderlin's theoretical writings. Chapter 1 attempts to establish this counter-position. Caution is necessary, as the argument for interruption can quickly become one that identifies such occurrences with novelty. Consequently, it is necessary to distinguish between arguments pertaining to the temporality proper to modernity and the temporality of fashion. Again, Benjamin's work is central, since the reference to the temporality of fashion refers to the object of critique—or at least one of the predominating objects of critique—within his writings on the nature of historical time. (This point is pursued in considerable detail in chapter 2.)

Benjamin is as much concerned to develop a critique of historicism as he is to engage critically with the problem of novelty. Novelty, as was suggested, can be understood as a form of interruption. And yet, were it to be allowed such a position, then novelty could be evoked in order to counter the expression of historical time as simple continuity; i.e., the temporality of historicism. It is a failure in this regard, failing because novelty is difference within variety rather than difference as a founding interruption. As a result, the need to differentiate between interruption ("awakening") and novelty is fundamental to any reworking of style. As will become clear in the investigation of Benjamin's critique of Jugendstil (ostensibly in Konvolut S of *The Arcades Project*—the task of chapter 3), novelty is not an interruption of continuity. On the contrary, it becomes one of the forms in which continuity appears. Once articulated in terms of novelty, continuity is then present as a series of different moments. Difference in this context is, therefore, no more than variety. Once this position is rearticulated in relation to commodity production, which in Benjamin's terms is an argument concerning commodity fetishism, novelty (coupled to the conception of historical time within which it can be positioned as an interruption) becomes the fetishization of the singular nature of the object. The object is, of course, not singular. Rather, singularity is the instant within the endlessness of variety. This is why Benjamin will identify this conception of the modern with the "time of hell," adding "the punishments of hell are always the newest things going on in this domain" (S1, 5).

This point can be taken a stage further. Novelty, while explicable in terms of singularity within variety—and as such it cannot be explicated as the individuation of the singular within difference—is also a conception of the new. Though only the new as appearance, since it is a conception of the new to the extent that the "new" figures within a formulation of historical time in which temporality is understood as simple continuity; i.e., time as linear progression. What this means is, of course, that to the extent that a distinction can be drawn between novelty—the interruption that figures within continuity and thus does not figure as an interruption as such—and a conception of interruption that is unthinkable within the conceptual constraints of historicism, then the new (or whatever term is going to designate this other conception) cannot be equated with mere appearance and therefore with style if the latter is reduced to no more than an appearance. However, this does not mean that it cannot be present in terms of its having a certain style. Only by allowing for the interarticulation of style and time can a distinction be drawn between simple novelty and the possible realization of an actual interruption. (Within the realm of art—in its widest extension—discerning this difference is one of the defining tasks of criticism.) The important point here, however, is that such a form of development—development as interruption—gives rise to a radically different sense of continuity. The difference is fundamental, since it is a continuity founded on the discontinuous and the different rather than one based on the endless moments of variety. (As has already been suggested, this endlessness is the temporality of fashion.) The conflict between variety and difference, in temporal terms between the discontinuous and the continuous, instantiates what will be developed throughout these essays; namely, the relationship between politics and time.[10]

The importance of Benjamin's critique of Jugendstil in this context is twofold. In the first instance, and in order that its full force be noted, it demands that the setting constructing the debate inaugurated by Heinrich Hübsch's 1828 pamphlet *In What Style Ought We to Build?* be taken into consideration. Such a setting positions Jugendstil as a moment already marked by the concerns of historical time. The severance of style and appearance is located in the formulation of the title's pamphlet in terms of a question. The use of the interrogative means that the way continuity is to be maintained at the present and sustained for the future now exists as a question. (The assumption of merely imitating the Greeks, the standard eighteenth-century response, evidenced in Winckelmann, for example, no longer holds.) The debate is as much about appearance becoming problematic as it is about the appropriate use of materials. Indeed, materials open up the other reason why Benjamin's critique of Jugendstil is of such significance. Jugendstil becomes therefore an overdetermined site. It is the moment in which style turns from its inherently public register and attempts to become singular and hence idiosyncratic. This negation of style in terms of singularity becomes its necessary reiteration as a commodity. In addition, with Jugendstil materials become stylized. Hence the question of materiality and the use of materials are central to its development. Benjamin is acutely aware of that centrality and it figures in his critical engagement. In part, the critique of Jugendstil reiterates the critique of novelty and thus the positioning of the object within the temporality of fashion. Once formulated within a setting in which materials and time are present, then, a concern with architectural materiality—what in the contemporary context should be taken, though only ever as a point of departure, to be a concern with tectonics—becomes political because of the interplay between what the potential of materials involves and the location of that potential within a conception of historical time that privileges discontinuity over continuity.

Developing a project that works through Walter Benjamin involves orienting that work in a particular direction.[11] In this context, this has meant distancing his concerns with the utopian and as a result repositioning the project in terms of the centrality of time. However, it is neither mere time nor just historical time. Time here has three determinations. It is a conception of temporality that incorporates the materiality of objects; it allows for the possibility of action because it thematizes interruption; and finally, in providing the basis for an analysis of the present and thereby allowing concerns proper to the present—a formulation posed at its most emphatic in Benjamin's "On the Concept of History"—it opens up the writing of history as determined by these concerns.[12] As such, a politics of time is not going to lie merely in the content. It lies as much in its own formal possibility.

Unforeseen Appearances

The essays in part 2 of this book start with Jože Plečnik and end with Édouard Manet. Given the problem of starting with architecture and finishing with art, moving from Plečnik to Manet may seem an extraordinary undertaking, in light of their relative chronological positioning. Plečnik's dates are 1872 to 1957, while Manet's are 1832 to 1883. However, these essays are centered on the question of place, expectations, and the political, and it will be argued that the nature of the differences between Plečnik and Manet provides a way of defining those concerns. In a sense, the project of all the essays in this part of the book can be succinctly encapsulated in Georges Bataille's description of Manet's painting *Olympia.* Bataille writes that she exists "no where" *(nulle part).* In being located beyond the oscillation between goddess and prostitute—and, to deploy Nietzsche's formulation, in being positioned beyond the convention of "good and evil"—the question of her place arises. In fact, as a question it has a far greater exigency than would be the case if the geographical correlations of good and evil were taken as delimiting her position. And yet this is no mere utopian gesture. The question is not of a speculative invention of another possibility. Rather, the task is one of reworking place beyond its usual determinations such that the same place becomes a site of alterity. The resultant effect is that what is meant by place is transformed in the process.

Plečnik's work, it will be argued, provides one of the last moments of symbolic architecture. Here particular attention will be paid to the facade of the National University Library in Ljubljana, Slovenia. The argument will be that what allows the incoherent presence of symbols to cohere is an organizing idea. The idea in question is a projective conception of Slovenian national identity. The undoing of this projection causes the symbols on the facade to no longer cohere. The facade therefore stages an understanding of place articulated within a synthetic formulation of national identity.

This understanding is precisely the one against which Nietzsche's joining of the modern and the "homeless" will be positioned. That positioning necessitates a conception of place located beyond the hold of any simple oscillation between the national and the international. The question of the symbol—appearance—arises, therefore, as the national and the international attempt to organize space in terms of symbols and modes of ornamentation. Any approach to a rethinking of place needs to start with the symbol. This mode of argumentation is developed in greater detail in chapter 4.

One of the most important arguments for symbols is that they envisage and demand a synthetic sense of community. The intention is that the monument or the memorial, for example, project and envisage a unified conception of community. The predicament of the modern means that this conception of unity is, in the present condition, either fleeting or merely pragmatic. The failure of the symbol to work in this inclusive way engenders two possible responses. The first is the nostalgic longing for a community. The response is to demand a renewing of symbols—or the attempt to make symbols contemporary—in order that a conception of a unified community could then be reactivated and renewed. This reactivation and envisaged renewal would allow for continuity. (Such a move can be futural, involving imagined communities that would, in a way to be determined, realize a latent potential. Equally, it can be steeped in the "past." The past in question would, of course, be a construct.)[13]

There is, however, a dramatically different possibility that demands having to work with the recognition that symbols have no place within modernity (other than as merely pragmatic gestures), because of their necessary connection to a synthetic conception of community.[14] This would be the case precisely because the concept of a synthetic community is antithetical to an affirmation of modernity. Note there is no argument here that this nostalgic moment—a moment bringing its own politics and forms of cultural practice into play—is not part of modernity. Indeed, it can be argued that part of what delimits the particularity of the modern is the conflict between attempts to reactivate synthetic conceptions of community (here the concerns of style and symbol, from a certain perspective, coincide) and the affirmation of the impossibility of this state of affairs.[15]

It may look as though all that is at work in a situation of this kind is the opposition between the national and the international, where the international would be that which denotes the collapse of community, while the national—whether as a given empirical reality or as a projected ideal—allows for synthesis. However, the opposition between the national and the international, while implicated within the problematic relationship between modernity and symbols, does not stage the position with sufficient acuity. It is not as though the countermove to national symbols—e.g., vernacular architecture—is a simple championing of the international. The latter, once articulated as a style, becomes nothing other than the symbolic expression of the value system of international capital.[16] As such, the move to the international cannot be readily differentiated from the internationalization of the symbols and brands that make up international capital. (This will be the case no matter how complex this setup will appear.) It should not be thought surprising that those symbols and thus those brands can come under attack. And this will be the case even if that attack is not undertaken in the name of either an exclusive sense of national identity (often defined in racial terms) or a form of religious fundamentalism. The choice given by the nature and components of this opposition has to be resisted. The national and the international form an either/or. Part of the argument of this project involves the claim that a significant element of the task of modernity is to think beyond, and thus move beyond, the complex determinations envisaged by any simple move between the national and the international. Rather than remain trapped within the structure created by this opposition, there needs to be a way of designating another possibility. Consequently, what has to be identified is a term which, while alluding to both determinations, is not structured by them. Moreover, it needs to be a term whose function will transform the ways identities such as the national and the international are understood. Even though it awaits argumentation, the *cosmopolitan* will be the term deployed to this end. This is a position developed in greater detail in chapters 4, 5, and 6.

Developing a conception of the cosmopolitan can be addressed as much on the level of style as it can on the perhaps more demanding level of a cosmopolitan sense of justice and hence in terms of another way of understanding international law (the international as the cosmopolitan). Here, rather than attempt this latter task it is the question of the symbol and thus of its relation to modernity—even if that relation becomes one marked by the symbol's own eventual impossibility—that will be the predominating concern. The topic in relation to which this concern will be pursued is architecture. Not architecture *tout court.* Rather, the way architectural concerns figure within a move from a conception of style as determined by a unifying and unified idea to the dispersal of that possibility within the cosmopolitan. (This will be developed as much by general forms of argumentation as it will by a consideration of specific architectural and urban projects.) The argument will work through the symbol, since it is the interplay of style and the symbol that allows for a critical engagement with the determining effects of historical time to be staged. Another task emerges therefore. In staying with the symbol, the problem to be addressed is the way style can be reintroduced such that it can be viewed as a temporal category and therefore one that opens the possibility for another politics of style and, in addition, another conception of the temporality of style. What will mark these "other possibilities" is that they will have resisted the temporality and the appearance of fashion. Implicitly, therefore, what will be reiterated in this context is Walter Benjamin's attempt to differentiate between the temporality of fashion and the temporality proper to an understanding of modernity thought in terms of a founding interruption. Equally, what also returns—and this return will continue to figure in all the chapters of this part of the book—is the question of appearance. Once the temporality of fashion is distanced, then what occurs at the same time is the distancing of the new—the new beyond the hold of novelty—which is explicable in terms of a "new" appearance.

A more sustained development of the cosmopolitan architecture is the subject matter of chapter 4. However, instead of concentrating on how the cosmopolitan is to be designed, emphasis will be given to the way an insistence on the cosmopolitan has a transformative effect on the nature of place. In straightforward terms, place is the site of architecture. Site can be understood literally as the location where the development or building is to be constructed. On a larger scale, place involves the national insofar as the place is the land of the nation-state. However, to the extent that a term such as the "cosmopolitan" can interrupt the opposition between the national and the international, then what occurs is a transformation of the way place is understood. This will have an effect—though not a deterministic effect—on the question of design. Indeed, the impossibility of determinism should be understood, once more, as evidence for the separation of style and appearance. The question of the appearance of cosmopolitan architecture exists as a question, one whose answer is continually being discovered.

As a prelude to the treatment of place, and though still integral to the development of the idea of the cosmopolitan within the realm of architecture, it will be essential to show in greater detail the link between historical time and the symbol. Again, time is fundamental. Once such a position can be demonstrated, then what becomes important is the way the cosmopolitan, in being positioned between the national and the international, interrupts their distinctive senses of continuity. Two components are integral to any definition of either national or international architecture. The first is that their presence as symbols brings with it the different sense of community evinced by such symbols. The second is that despite the literal place in which such architectural objects are to be found—and by "literal place" what is meant could be as straightforward as the coordinates on a grid—what has to be taken up in relation to this connection of symbol and place is that a shift in the nature of the symbol brings with it a related shift in how place is to be understood. In other words, part of the consequence of this move is that a reconfiguration of the symbolic results in the need to rethink place. This will be the case even if the place in question is identical—literally the same geographical location. Within this rethinking, it would follow that precisely because the place is on one level identical, that would allow it to emerge as the site of difference. Difference would mark the traces of a history marked by conflict: traces that would have to be eliminated in any conception of place that reduces it to the status of an uncontested site.

Within the discourse of architecture, one of the most important attempts to develop an understanding of place that resists the twofold determination of the national and the international has been Kenneth Frampton's concept of "critical regionalism."[17] The final part of chapter 4 will therefore develop a conception of the place of the cosmopolitan by working through the important insights afforded by Frampton's argument. What will emerge is that a consideration of place—a consideration opened up by the question of style—will occasion a rethinking of place that can be shown to have an inherently political dimension.

As has already been indicated, chapter 5 will start with an analysis of the work of the Slovenian architect Jože Plečnik. Due to his use of symbols, Plečnik's work was until recently considered to be postmodern architecture *avant la lettre.* Not only is such a position false—since it is based on a systematic misunderstanding of the use of symbols—but accepting it would give too great a significance to postmodernism as an argument about the nature of historical time. The assumption here is that postmodern architecture is a specific development within the history of decoration and as such it does not come after modern architecture in any real sense. The consequence of working with this assumption is that it reopens the question of what would count as modern architecture. The force of this question is that it is concerned as much with questions having to do with style—and thus the appearance of the modern—as it is with the nature of historical time. Such an opening provides the locus in which to develop the idea of a cosmopolitan architecture. It is important to note that this will not occur in the abstract. The strategy will be to work through specific architectural and urban projects. In regard to Plečnik, emphasis will be given to his National University Library in Ljubljana, Slovenia. Consideration will also be given to projects by Ashton Raggatt MacDougall (the National Museum of Australia in Canberra), Herzog and de Meuron (the library at Eberswalde University in Eberswalde, Germany), and LAB Architectural Studio (Federation Square in Melbourne, Australia), among others.

The interrelationship between style, place, and the cosmopolitan—all of which are elements figuring in any delimitation of the site of style—will allow for a rethinking of movement backward and forward across borders. The significance of the border is that it is taken to define the determination of place. While borders may be arbitrary, they delimit the complex in which law operates. The border dramatizes the link between a nation's self-conception of its own identity—the problematic status of the latter enacted by the impossible possibility of the symbol—and the law. These considerations give the border, understood as an edge condition of the nation, greater exigency. Troubling this edge is the refugee. In this context, what this means is that a politics of place will allow for an approach to the urgent issue of the refugee. With the refugees there is a situation that has a daily and ineliminable insistence. Part of that insistent quality is the difficulty of formulating a political response that falls outside the hold of a simple moralism. Part of the argument of chapter 6 starts from the position that instead of viewing the refugee as the exception and allowing that exceptional status to have a determining effect, another approach is necessary. In fact, once there is a repositioning of emphasis in which the urban and the social are defined in terms of movement rather than stasis, the position of the refugee emerges as the "norm." The question, then, concerns how this "norm" is to be understood. Beginning to answer it demands continuing to rework terms such as "hospitality," "acceptance," and "place," since they mark the location and conditions of reception. What will occasion such a reworking, in this context, is the argument already developed concerning the dispersal of the effect of the symbol within the cosmopolitan. As such, this chapter will take up and reposition the arguments presented through the book as a whole. Any consideration of place—whether that consideration is driven by political concerns or the apparently neutral concerns of policy and planning—either affirms or denies this reworking of place in terms of its inherent complexity. Here, again, there is an echo of Walter Benjamin's evocation of awakening.

As has been argued, awakening interrupts continuity. To reiterate the argument: here the continuity in question is twofold. In the first instance, it is the assumption that the position of the refugee is explicable in terms of the conceptions of borders and border relations defined by the exclusivity of inside/outside relations—relations in which the identity of each element is already determined in advance and has an essential quality. Second, it is that the policy and practices linked to the acceptance of these positions, and therefore of the conceptions of identity that come with them, define the realm of the political. The acceptance or rejection of these positions becomes a simple either/or. Again, interruption will involve refusing the hold of this either/or. What will be argued is that posed against both these possibilities are actions arising from the recognition that simple continuity amounts to no more than the acceptance of a setup that is integral to the replication of a series of positions having no basis outside the relations of power that they sanction. Hence, what they demand is their own self-replication. Awakening to this situation is not just a realization. It stems from, while enjoining, action.

In more general terms, this project brings together differing forms of analysis with differing effects. Taken as a whole, it will allow for a way of construing the nexus of style and time such that it yields differing modes of inquiry. In the end, however, an inquiry of this nature is not removed from actuality; it is intended to have an impact on the formulation of policy and practice. This is not to argue that these essays have an instrumental quality. Rather, the reformulations of "place" and the cosmopolitan—and allowing architectural concerns a place within them—bring specific openings into play. In the first opening, what is allowed is the possibility of an affirmation of discontinuity and thus of strategies of interruption. However, this is not simply a speculative move, let alone a utopian gesture. It needs to be assumed that a certain reality can be given to the possibility of maintaining that which is discontinuous in relation to an enforced and enforcing continuity. That reality is practice. Specifically, it will be the decision to opt for one form of practice rather than another. It will be a decision in which difference is fundamental. The second opening—the temporality of the modern—is that this possibility, the reality occasioned by this type of affirmation, is associated with arguments about the nature of modernity. Again, there will be a type of necessity insofar as modernity will need to be understood as a discontinuity that awaits its own affirmation. The third opening—the symbol—is that bound up with the first two openings is the problem posed by the symbol understood as already interarticulated with arguments concerning the relationship between community and historical time. As will be seen, the problem of the symbol rehearses the questions that are already present and thus operative within any consideration of the new and its differentiation from both the form and the temporality of novelty; in sum, it rehearses the question of appearance. The complex interrelationship between these three elements will provide part of the overall continuity of the book.

The conclusion to be drawn from these essays will be that understanding the specificity of the present cannot occur either by the simple evocation of history or by positing the unproblematic presence of the political. In both instances, the historico-political question is how these terms are to be understood. The argument advanced here is that central to developing such an understanding is the recognition that time plays a fundamental organizing role within them. Access to time—to time as it figures in the questions of history and politics—has to be found within the continual intrusion of temporality into the public sphere. That intrusion is neither abstract nor is its presence predetermined. This is the mark of modernity's ineliminability. That intrusion is the question of style.

Style and Time

Part One : **Working through Walter Benjamin**

1. Benjamin's Modernity

Any argument that starts with the claim that it concerns a theory of modernity is constrained to account for the nature of modernity's inception. Even in working with the assumption of modernity's presence, there would still have to be a description of that which was located in its differentiation from the modern. Part of the argument to be developed here is that for the major thinkers of modernity, its occurrence is thought in terms of a break or an interruption. In part 1 of this book, the particular project is to locate that thinking in the writings of Walter Benjamin. A context therefore is set by those writings and the presence within them of attempts to develop a relationship between modernity and its necessary interarticulation with a philosophical conception of historical time. Given this context, the opening question has to concern the specificity of interruption within those writings.

How is interruption to be thought? What is the conception of interruption at work within Benjamin's writings? Although it appears as a motif in his engagement with Romanticism and is then repositioned—if not reworked in the later writings in terms of a thinking of historical time—interruption as a mode of thought within Benjamin's work can be identified under a number of different headings.[1] In each instance, what insists is the question of what interruption stages. In Benjamin, as will become clear, interruption is the term through which a theory of modernity can be thought. This is not to argue that it is identical with the conception of modernity located in Benjamin's writings as such. Rather, it is modernity as an interruption, one that has to be maintained and which will vanish within the resurgence of historicism understood as the insistence of continuity in the face of discontinuity, that marks the move from a specifically Romantic motif to a thinking of historical time. The Romantic motif of interruption provides a possible form for such a thinking of historical time. The direct consequence of this is that to the extent that this latter point is the case, then a theory of modernity will owe as much to a Romantic heritage as it will to one coming from the Enlightenment.[2] Indeed, it can be further argued that thinking the particularity of modernity as an interruption depends upon the successful distancing of the conception of historical time within the Enlightenment tradition.

Interruption is named in different ways. Perhaps the most emphatic, and the one that will allow this theme to be traced here, is the "caesura." The aim of this chapter is to develop an understanding of interruption in terms of the "caesura," and then to note the effective presence of this specific mode of thought within a number of different texts. Often interruption will be named differently. Rather than attempt a synoptic exercise, two particular moments will be taken up. The first concerns the work of the caesura in Benjamin's essay "Goethe's Elective Affinities," and the second is the recurrence of the term in Konvolut N of *The Arcades Project* (N10a, 3). In regard to Benjamin's own chronology, these texts mark the beginning and the end of his writing career. While Benjamin wrote both his doctoral dissertation and his essay on Hölderlin prior to the Goethe essay, the latter can be seen as the point of departure for both the concept of criticism he developed in the dissertation and his sustained engagement with the Romantic heritage. *The Arcades Project,* while not finished in a literal sense, always brought with it the possibility of never being finished. As such it was the work that truly marked the end of Benjamin's writings.

Romantic Interruption

Almost at the end of Benjamin's extraordinary study of Goethe's novel, he writes that a particular sentence contains what he describes as the "caesura of the work." Analyzing this claim will open up the way the caesura is staged in his early writings. The passage in question is the following:

> In the symbol of the star, the hope that Goethe had to conceive for the lovers had once appeared to him. That sentence, which so to speak with Hölderlin contains the caesura of the work and in which, while the embracing lovers seal their fate, everything pauses, reads: "Hope shot across

> the sky above their heads like a falling star." They are unaware of it, of course. (*SW* 1:354–55/*GS* 1.1:199–200)

The presence of the star cannot be divorced from its presence as a symbol. The text is clear: "Denn unter dem Symbol des Sterns" ("In the symbol of a star"). Introduced with the symbol is the split that works within the caesura and which is registered in the lovers' nonregistration of the star as the symbol of hope. Understanding that split means paying attention to the complex relationship between time and the Absolute as it figures in the symbol insofar as the symbol is evidenced in this passage. (At this stage in Benjamin's development, he is yet to formulate a sustained distinction between symbol and allegory.) Benjamin has allowed here for a conception of the symbol that departs from the simultaneity of the relation between symbol and the symbolized, though it equally departs from the hermeneutic demands of surface/depth as the setup through which the symbol is constrained to be interpreted. The opening up of the symbol occurs within what could be described as a destruction entailing ontological and temporal considerations. Destruction figures in the Goethe essay in a number of different places. One of the more significant is in terms of the "torso."

Benjamin refers both in the Goethe essay and in the doctoral dissertation to the "torso." In the case of the dissertation, the term is used to argue that the particular "can never coalesce with the Ideal" but has to remain "als Vorbild" (as a prototype). In the Goethe essay the symbol is also linked to the "torso." It is presented in relation to the work of "the expressionless." Benjamin writes: "Only the expressionless completes the work by shattering it into a thing of shards, into a fragment of the true world, into a torso of a symbol" (*SW* 1:340/*GS* 1.1:181). What is a "torso of a symbol"? The first part of the answer to this question is that it is a result: the consequence of the work of the "expressionless." The work is completed in its being fragmented. The mistake would be to read this as a literal claim. There aren't any shards; there will not have been any fragments. Rather, the moment (and it is a moment, Benjamin writes, *in einem Augenblick*) is that in which the most severe form of irreconcilability occurs. The torso of a symbol, however, is not given within the structure of necessity demanded by diremption, since it does not envisage its own overcoming or resolution. Rather, it is the staging of an opening that can only ever be maintained as this opening. Being maintained in this manner, it defines a predicament in which the problem of closure and thus resolution is staged without an end being envisaged.

What then of the "torso" in this predicament? As a torso, the symbol has been stripped of the structure and thus of the possibility of temporal simultaneity; nonetheless, this cannot be interpreted as opening up a field of infinite deferral. The work is still completed. The expressionless completes. Again the text is clear: Benjamin states, "vollendet das Werk" (completes the work). It is completed by the occurrence within it that is the work of a temporal register that cannot be assimilated to the temporality of expression. What this means is that what completes the work is integral to the work's formal presence and not to the "content" of its narrative. The "expressionless" is not the interruption of continuity, nor is it simple discontinuity. It completes the work by showing, on the one hand, the perpetual vacuity of expression if expression were thought to voice the all; and on the other, by demanding of the work that it recall—recall within and as its work—its separation from the eternal. While more needs to be said, the introduction of time allows the problem of the nature of the caesura, and in this context its relation to hope, to be staged. In the passage already noted the caesura enters with a particular purpose. The expressionless understood as "a category of language and art"—though not of a work or genre—"can be no more rigorously defined than through a passage from Hölderlin's 'Remarks on Oedipus'" (*SW* 1:340/*GS* 1.1:181), to which Benjamin adds that the deployment of the caesura beyond its use in a theory of tragedy has not been noticed, let alone pursued with adequate rigor.[3]

Two points therefore arise. The first is that the caesura allows for a rigorous definition of the expressionless. Second, the caesura is to be used other than in its employment within a theory of tragedy. The caesura is precisely not an emblem of rhetoric. On one level the caesura and the expressionless are different names for the same possibility, namely, an interruption that yields completion. It is this possibility that needs to be pursued by a return to the passage in which the completion of Goethe's novel is identified as occurring in a single sentence. How could it be that a sentence might "contain the caesura of the work"? What is shattered in this case? Where are the shards? Here there are no twitching limbs vainly gesturing at what remains, i.e., the torso. How then is this claim to be understood? Moreover, the passage in which this phrase—"which will complete the work"—is presented does not occur at the novel's completion. It may set the seal for what will occur and yet it occurs pages from the end. How then does it work to complete the work? For Benjamin this has to be the question proper to criticism, if only because the answer would "provide detailed knowledge of the work."

It is essential therefore to return to one of Benjamin's formulations of criticism. Only with an understanding of criticism will it become possible to follow the role attributed to the caesura in the Goethe essay. The essay is, after all, a work of "criticism." The passage in question moves criticism through a number of vital stages. While the passage is detailed—containing in addition an important reference to Schlegel's own criticism of Goethe's *Wilhelm Meister*—its detail is essential:

> The legitimization of criticism—which is not to posit criticism as an objective court of judgment on all poetic production—consists in its prosaic nature. Criticism is the preparation of the prosaic kernel in every work. In this, the concept of "preparation" is understood in the chemical sense, as the generation of a substance through a determinate process to which other substances are submitted. This is what Schlegel means when he says of *Wilhelm Meister:* "the work not only judges itself, it prepares itself." The prosaic is grasped by criticism in both of its meanings: in its literal meaning through the form of expression, as criticism expresses itself in prose; in its figurative meaning through criticism's object, which is the eternal sober continuance of the work. This criticism, as process and as product, is a necessary function of the classical work. (*SW* 1:178/*GS* 1.1:109)

Criticism is that approach to the work in which the identification of its particularity allows for its incorporation into what Romanticism would have identified as "the realm of the Absolute." The move, in the most direct sense, would be from the "prosaic kernel" to the prose of criticism. The extent to which a work is criticizable is the extent to which it prepares itself (is prepared) for this possibility. The complicating factor in this passage is how the distinction between the "literal" and the "figural" is to be understood. For Benjamin, "prosaic" has two meanings. The first refers to its presence defined within the context of the passage as "unmetrical language," i.e., the prosaic expressed in the prose of criticism. However, the prosaic is also "grasped by criticism in a figurative sense" as "the eternal continuation of the work." What this means is that criticism holds to particularity while, at the same time, allowing for the particular's absorption into the Absolute.

Criticism is able to allow for the completion of the particular work to the extent that the work is criticizable. As it is formulated in the Goethe essay, this signals the presence of the possibility of showing "in the work of art the virtual possibility of formulating the work's truth content" as the "highest philosophical problem." The latter is of course the staging of the Absolute and its impossible possibility. The moment that brings this together is the caesura. As has already been intimated, the first reference in Benjamin's text to this term that is worth noting concerns his identification of the caesura as it figures in Hölderlin's "Remarks on Oedipus." It is important to return to the actual text he cites. The Hölderlin text as cited by Benjamin is as follows:

> For the tragic transport is the actually empty and the least restrained.—Thereby in the rhythmic sequence of the representations wherein the transport presents itself, there becomes necessary what in the poetic metre is called caesura, the pure world, the counter-rhythmic rupture—namely, in order to meet the onrushing change of representations at its highest point, in such a manner that not the change of representation but the representation itself soon appears. (*SW* 1:340–41/*GS* 1.1:181–82)

Hölderlin's formulation is more complex than suggesting a form of interruption that would only ever be a counter-rhythm. Meter does not measure the interruption. That would make the caesura a literal breaking apart. Rather, such a rupture must take place on the level of representation and presentation. The site of interruption is the "sequence of the representations" and their movement is that of the "onrushing change of representations." The sequence and the movement produce the site of interruption. This sequence cannot be straightforwardly conflated with plot. Sequence and movement need to be viewed in temporal terms. They involve a particular form of unfolding: one which articulates a sequential temporality. The caesura is positioned by place—insofar as it can be located—while it is not the work of place. Thus, it is not another occurrence. The complicating factor here is that interruption is the interruption of a certain temporal sequence, though it is equally the interruption of the possibility of reading that sequence as the unfolding of the purely transcendental. In other words, the work is neither regulated nor caused by that which is external to it. The former element is the one that comes to dominate Benjamin's later writings. Nonetheless, the other element is important, since what is refused by it is the possibility of an eternal other, either as God, idea, or myth, to provide the artwork with its legitimacy and, though this is probably to reiterate the impossibility of legitimacy, to offer the locus and thus determine the nature of critique. Critique is not that which occurs in terms of a relation between an external element that causes the internal components to have their specific determinations.

What remains elusive in this presentation of the caesura—and here it cannot be restricted to the caesura, since it involves the other forms of interruption—is how such an event can "give free reign to an expressionless power inside all artistic media." The answer to this question is there in the almost possible object of attainment identified by the use of the term "sobriety." It marks the point of connection between measure and the measureless. As such it is the return of the problem of particularity. This time, however, it is posed in a different way. Rather than the particular, it is the Absolute that has centrality. The Absolute here is marked by an impossible possibility. At the same time it is also generative. However, despite being productive, the Absolute cannot be produced. It can neither be made nor shown. Read back through the caesura—and while not wishing, again, to conflate them—it marks the interruption that yields an artwork. It presents that which is proper to art. This is the "expressionless power inside all artistic media"; i.e., the Absolute. The Absolute, the nature of its presence, already turning within semblance, cannot be reduced to that to which "mere semblance" gestures.

How is semblance to be delimited? In a discussion of "Goethean figures" and thus as an integral part of the work's critique, Benjamin, drawing on the critical apparatus he had already established, writes of those figures:

> [They] can appear to be not created or purely constructed but conjured. Precisely from this stems the kind of obscurity that is foreign to works of art and that can be fathomed only by someone who recognizes its essence in semblance. For semblance in this poetic work is not so much represented as it is in the poetic representation itself. It is only for this reason that the semblance can mean so much; and only for this reason that the representation means so much. (*SW* 1:345/*GS* 1.1:187)

At work in this formulation is that which arises from the operation of critique. In the first instance, there is an appearance of figures having one source rather than another. Here again, the detail is necessary. The formulation is precise. Goethean figures "can appear" to have arisen through an act of conjuring and if that were the case, the critique of conjuring as "having nothing in common" (*SW* 1:340/*GS* 1.1:180) with the generation of art would have been rendered otiose. The problem of this appearing is the problem inherent in the work. Its presence attests to the necessity of critique, and thus to critique as an activity done in relation to a work that sanctions it. Rather than taking what appears as appearance, in the end mere appearance, the reverse situation needs to occur. There has to be the recognition of what is essential to art in the semblance; the "Wesen" in the "Schein." With the *Elective Affinities,* and it should be noted that Benjamin specifies that in this work there is not a "presentation of semblance," it is the presentation itself. This is the reason why semblance can have the meaning that it does, and reciprocally this is why the presentation itself is imbued with such meaning. Again, it is essential to see what is being distanced. Not only is there a sustained refusal to interpret appearance as representational and therefore as standing for something other than itself; there is also a distancing of the possibility that semblance acts out what it can only gesture at in action without being such an action.

Understanding the import of this claim concerning the presence of semblance depends upon accepting Benjamin's identification of the two elements that determine the interpretation. The first is that "the subject of the *Elective Affinities* is not marriage," and the second is that "belief in Ottilie's beauty is the fundamental condition for engagement with the novel" (*SW* 1:338/*GS* 1.1:178–79). This is not "the appearance of the beautiful"; rather, it is the "semblance-like beauty" that is central. This shift has to be recognized as a move from content to truth; that is, from a concern with the "material content" to a concern with "truth content." It is not as though marriage and the concerns of bourgeois gentility are absent. Rather, they only figure within the work of truth. Before pursuing this move to semblance and thus to the complexity surrounding semblance, it is essential to note that the emergence of beauty occurs as part of the process of critique. Underlining the importance of this shift is not, therefore, a mere passing remark. It delineates how Benjamin's essay is also a work of critique.

Once critique is linked to an engagement with truth rather than content, then content has to be repositioned in relation to truth. The detail of Benjamin's engagement with the constitutive elements of the novel is, though only in this instance, not necessary. What has to be retained, however, is the direction of that engagement. Benjamin's move is to reposition semblance and thus give it its full philosophical force. Semblance opens up the realm of the Absolute and the relationship between the particular and the Absolute once it is understood that semblance stages both itself and the Absolute. (The presence of the latter is to be thought in terms of the impossible possibility of the Absolute's presence.) Precision is essential here. Benjamin's claim is that critique works within the opening between particular and Absolute. It needs both elements—particular/Absolute—since the interrelationship of these elements comprises the work of art. This is why in regard to the treatment of beauty in Goethe's novel, in which beauty becomes "the object in its veil," Benjamin will write that:

> The task of criticism is not to lift the veil but rather, through the most precise knowledge of it as a veil, to raise itself for the first time to the true view of the beautiful. To the view that will never open itself up to so-called empathy and will only imperfectly open itself to a purer contemplation of the naive: to the view of the beautiful of that which is secret. Never yet has a true work of art been grasped other than when it ineluctably represented itself as a secret. (*SW* 1:351/*GS* 1.1:195)

The opening line contains the key to this passage. There would seem to be a twofold possibility. The first links criticism to the process of revelation and thus the uncovering of an inner truth. This is precluded, since criticism is not concerned with lifting the veil. In the same way, a fetishism of the veil generating an interpretive mysticism would still attribute to the veil the quality of harboring depth. This would imply that the veil is literally the veil. Both these possibilities are curtailed since criticism is already informed. Benjamin is clear: the veil is known. It is an object of knowledge. Intuition or empathy would fail to interrupt the work of the infinite. Knowledge rehearses the petrification of the object; the object of knowledge. Knowledge does not provide access to the secret. Knowledge is knowledge of the artwork as the secret. Knowledge maintains the secret, though as known. The limit is established by the effective nature of the Absolute. While accounting for presence—and allowing for its present incorporation as part of the particular's presence—it can never be present as itself.

If there is a way of generalizing what is at work in the complex relationship between interruption and criticism, then it can be captured in the claim that what the caesura allows is the relationship between the particular and the Absolute to be thought. In doing this the artful nature of the artwork is presented. Criticism in the context cannot be thought other than in its relationship to the work of the Absolute. The complex presence of the Absolute and the way it figures within, if not providing the very ground of, Benjamin's engagement with Early Romanticism opens up the move to his later concerns with history. That concern is not with the detail of history—Rankean "facts"—but with the temporality that such facts display and within which such facts are able to be displayed. History cannot be thought other than as a philosophy of time.

History: Interruption

In moving from a concern with criticism to the concerns of Benjamin's *Arcades Project,* the difficulty any commentator faces is how to account for the repositioning. Perhaps the key interpretive question is: Is there a retention of the Romantic conception of the Absolute (in sum, a conception in which a particular work is both itself and the Absolute at the same time)? Prior to any attempt to answer this question, what has to be addressed is the move from interruption in the writings directly concerned with Romanticism to a more generalized sense of interruption. Prior to turning to the passage from *The Arcades Project* in which the term "caesura" figures, two specific formulations of interruption need to be noted. The first comes from "The Work of Art in the Age of Its Technological Reproducibility" and the second from "On the Concept of History."

As a text, Benjamin's essay "The Work of Art in the Age of Its Technological Reproducibility" is full of remarkable moments, shock insights that attest to the interruption that yields the work of art within modernity. One of the most emphatic occurs in the following passage:

> Let us assume that an actor is supposed to be startled by a knock at the door. If his reaction is not satisfactory, the director can resort to an expedient: he could have a shot fired without warning behind the actor's back on some other occasion when he happens to be in the studio. The actor's frightened reaction at that moment could be recorded and then edited into the film. Nothing shows more graphically that art has left the realm of "beautiful semblance" which for so long was regarded as the only sphere in which it could thrive. (*SW* 4:261/*GS* 1.2:491)

What is this "beautiful semblance" where art was thought to "thrive," and in which it can "thrive" no longer? What type of change has occurred such that this dislocation and thus subsequent relocation have come to pass? The reference made in this 1936 text is both to the early Romantics (and thus to Benjamin's own engagement with that heritage) and to a sustained engagement with the topic of beauty that reappears throughout his work. A significant instance of that engagement is the long footnote on beauty in the essay "On Some Motifs in Baudelaire" (*SW* 4:352/*GS* 1.2:638–39). This footnote signals a historicization of beauty that was not as evident in his earlier writings. As has already been intimated, the most sustained discussion of beauty as "beautiful semblance" appears in the final section of his essay "Goethe's Elective Affinities." In the final pages of that text, Benjamin introduces—perhaps reintroduces—the task of criticism in relation to beauty. This relation is central. All of Benjamin's work on art has been concerned with detailing the task of criticism. Criticism is the key to the doctoral dissertation. Indeed, the complex relationship between philosophy and criticism is evident in the opening line of the prologue to *The Origin of German Tragic Drama* (*OT* 27–56/*GS* 1.1:207–37). In the published version there is an important modification of the earlier drafts. The change introduces the problem of philosophical style by inscribing the problem of presentation into the presentation of philosophy itself. The first passage is from the draft and the second is from the published version.

> Es ist der philosophischen Erkenntnis eigen, mit jeder Wendung von neuem vor der Frage der Darstellung zu stehen. [It is characteristic of philosophical knowledge that it must continually confront the question of presentation.] (*GS* 1.3:840)

> Es ist der philosophischen *Schrifttum* eigen, mit jeder Wendung von neuem vor der Frage der Darstellung zu stehen. [It is characteristic of philosophical *writing* that it must continually confront the question of presentation.] (*OT* 27/*GS* 1.1:207; emphasis mine)

While the shift from "knowledge" to "writing" is of great significance in terms of the development of Benjamin's text, what is interesting for these concerns is that writing becomes a practice stemming from a particular formulation of philosophical activity. The initial use of the term "knowledge" *(Erkenntnis)* creates the link to criticism, for it is at the end of the Goethe essay that criticism and knowledge are interconnected. What is important in the move from Benjamin's early texts to the later ones is that the conception of art is inextricably bound up with the task of criticism once criticism is defined in relation to knowledge. A shift in the nature of art enjoins a concomitant shift in the activity of criticism and thus of the philosophy of art. While that shift occurs, what is not lost is the link to knowledge.

Does this mean, however, that the shift from the "realm of beautiful semblance" detailed in the passage cited above is at the same time a move away from a thinking of art conditioned by the Absolute? Drawing such a conclusion would be too hasty. Clearly, what Benjamin can be interpreted as suggesting is not that the Absolute no longer figures in how the work of art is to be understood, but that the locus of art and thus what counts as art's work has changed. The move from the identification of art with poetry—or at least if not with poetry, then with literature in the broadest sense of the term—and thus the capacity to generalize about art based on that identification, has ceded its place to a definition of art in terms of what produces it. In regard to the work of art, what is occurring can be reformulated as a move from *poesis* to *techne.* Even in allowing for this reformulation, the question that returns is the extent to which such a move rids itself of the Absolute. And yet this question cannot be posed as though the answer were all or nothing. It is more likely the case that in the move from one to the other—*poesis* to *techne*—the Absolute rather than vanishing comes to be redefined. While art will continue to be defined via the activity of criticism, understood either implicitly or explicitly, in relation to the Absolute, the shift of the content of that definition will yield a differing understanding of the Absolute.[4]

As a generalization, the contrast is between two different possibilities for art and criticism. The nature of that difference is to be understood in terms of the relationship between time and the object. *Poesis* involves a different relationship than the one at work in art defined as *techne.* Indeed, it is because the relationship is formulated in this way that the temporal considerations at work in the latter—the conception of the work of art determined by *techne*—are such that they open up as historical concerns. Not the concerns of history as such, but in terms of the temporality proper to that conception of history that is constrained to undo the identification of history with the temporality of historicism. (The latter being the temporality of continuity that is sustained either in terms of simple chronology or in terms of the endurance through time of concepts—for example: beauty, genius—that are taken never to change.) Both the need for, as well as another sense of interruption, occur at this precise point. Prior to looking at passages from "On the Concept of History" and *The Arcades Project* in which the conception of interruption figures—and in the case of the latter is identified by the use of the term "caesura"—it is essential to note, if only in passing, the nature of the shift.

What determines Benjamin's initial sense of interruption is the necessity that the activity be internal to the work. The work "prepares" itself to be criticized. There is an extent to which the work has an autotelic nature. The link between the work of the "expressionless" and the activity of criticism—both are involved in differing forms of the work's ruination—is that they are defined in relation to an activity that originates in the place and presence of the Absolute. Once a work can be construed as criticizable, there is little that stands in the way of the practice of criticism. While it is true that Benjamin has harsh words to say concerning Gundolf's interpretation of Goethe, and while he shows a growing awareness of the political nature of criticism, it remains the case that there is no theorization of that which stands in the way of the activity of criticism. In other words, it is not the case in the early writings that there is any recognition of the need for a preliminary move, one which would allow for criticism. Such a move is not preliminary in the sense that it is prior to the activity of criticism. Rather, criticism means dealing with the way in which a given work of art functions as a work of art while also stripping that work of its insertion into the temporality of continuity—what Benjamin identifies as "historicism"—and thus disrupting the interruptive structures that accompany that hold.

Two points need to be made here. The first is that what the identification of the possibility of inserting, or cutting, a segment into a work—a possibility signaled in the citation from "The Work of Art in the Age of Its Technological Reproducibility" given above—indicates is that there is a shift in how the art object is understood. Part of this change is that the way the work of art works as criticizable changes. The second and related point is that the possibility of the work's absorption into the temporality of continuity is now a possibility that is inherent in the work itself. In the shift from *poesis* to *techne,* the work of art does not prepare itself to be identified as something particular. In the move from *poesis* something else takes place. Henceforth, the work of art is always prepared for its absorption into the realm of continuity. As has been indicated, what this means is that the activity of criticism—and perhaps it is possible to go further and argue that this is the task of the progressive critic—necessitates the interruption of that enforcing continuity; an enforcing that is inherent in the technical nature of the object.

Benjamin identifies the problem of historicism—understood as the temporality of sequential continuity—in the following passage from "On the Concept of History." Of central importance in this passage is the use of the imagery of the rosary beads. It provides a clear example of the way continuity has to be interrupted in order for the potential within and for art to be released. Of equal importance is that instead of writing about the critic, Benjamin will now write of the "historian." The passage in question is the following:

> Historicism contents itself with establishing a casual connection between various moments in history. But no fact that is a cause is for that very reason historical. It becomes historical, posthumously, as it were,

> through events that may be separated from it by thousands of years. A historian who takes this as his point of departure stops telling the sequence of events like the beads of a rosary. Instead, he grasps the constellation which his own era has formed with a definite earlier one. Thus he establishes a conception of the present as the "time of the now" which is shot through with chips of Messianic time. (*SW* 4:397/ *GS* 1.2:704)

A beginning can be made here with the interruption signaled by the use of the rosary. The question has to be: what does it mean to "stop telling the sequence of events"? Here there is a decisive formulation of interruption. And yet within the formulation, what is interrupted has a more complex quality. It is the projection of unity or synthesis—or what Benjamin will identify elsewhere as "universal history," having Kant rather than Hegel in mind—that has to be undone. No longer is this destructive move made in the name of anything other than an intervention within temporal continuity. Precisely because it is an interruption that involves a specific orientation that can be as much philosophical as it is artistic, the demands of that orientation, itself demanding a decision, allow what is taking place an inherently political dimension; again, what is involved is a politics of time.

Interruption as a figure within Benjamin's writings is linked to the dominance of historicism. Again, this is not a simple conception of the historical and thus of historical time. As has already been argued, what takes place within historicism is the naturalization of chronology on the one hand, and the naturalization of myth on the other. Working within both is a continuity that effaces the question of whose history is being told or narrated, and thus for whom and for what end a given history is being constructed. The act that denaturalizes both myth and chronology is the interruption. The immediate consequence of this interruption is the reconfiguring of the present. With that reconfiguration the present emerges as the "now"—a temporalized and historicized now—that generates the nature of the philosophical and therefore, and at the same time, the political task.[5] What this means is that in Benjamin's later writings, a twofold register is added to the locus of interruption. In the case of the earlier work, the locus was the work of art itself. Marking the move is the incorporation of the work of art into a time of the present in which whatever determines the work's specificity can be effaced. Effacing specificity occurs because what marks the work is its capacity to interrupt the time of the present. This interruption occurs as long as the temporality of the present is thought in terms of continuity. What this means, in addition, is that the present is not thought outside its insertion into continuity. This accounts for why Benjamin argues for the urgency of making something a concern for the present. If this does not occur, then the present does no more than form part of the "appearance of permanence" (N19, 1).[6] As such, the present is lost from the present. Thinking the present is already an interruption; an interruption yielding the present released from the hold of appearance. In part, what is repeated here is the structure of thought that has already emerged in Benjamin's identification of the task of criticism. Fundamental to that identification was the recognition that the truth of appearance does not lie beneath the surface. It is neither hidden nor present as an organizing deep structure. (Again, this is Benjamin's objection to Hegel—an objection already evident in his use of the term "cunning"—for whom such an organization level is essential.) For criticism, what is involved is developing a detailed knowledge of appearance, and such an approach puts in abeyance any residual utility that the opposition surface/depth may have had. Furthermore, on demanding that centrality be given to the particular work, that demand not only repositions how appearance is to be understood; it does so on the condition that an act of interruption yield the work as an object of criticism. This is not to argue that exactly the same approach occurs with the historian; nonetheless, the importance of the affinity remains.

The interruption occurs when the historian stops "telling the sequence of events like the beads of a rosary." Here there is a decision that interrupts. This position is made possible by a shift that can be traced from the work of art defined in terms of "beautiful semblance" to the artwork's inescapable connection to reproducibility and thus to technology. The methodological consequences of this interruption redefine how destruction and therefore ruination are to be understood. This other possibility is signaled in *The Arcades Project* as follows: "Historical materialism has to abandon the epic element in history. It blasts the epoch out of the reified 'continuity of history.' It also blasts open the homogeneity of the epoch. It saturates it with ecrasite, i.e., the present [*Gegenwart*]" (N9a, 6; trans. modified). Signaled in this passage is the decision to abandon continuity. That abandoning allows for, and at the same time is the opening up, of the epoch's homogeneity. What is meant by epoch is rescued and transformed in the process. The blasting open allows the fallout to contain the elements of historical work. The position being staged here needs to be run both ways. In the first instance, it has to be argued that in the detritus of history—what has been cast out of epic history—there lies the potential to interrupt continuity. Continuity may have been founded on just such an elimination. In the second instance, it is by blasting apart continuity that what looks to be insignificant, or merely awaits incorporation into a form of continuity or totality, may contain the potential to redefine the present and, more significantly, to have consequences that are potentially as much political as they are philosophical or historical. Occurring in this process is an act of rescue in which images of the past have a capacity to define the present. The rescue is the release—or attempted release—of that potential. In defining it—and again it should be remembered that such an act of definition is the result of a decision—the present comes to be established in contradistinction to the present of continuity. In the formulation of "On the Concept of History," "every image of the past that is not recognized by the present as one of its own concerns threatens to disappear irretrievably" (*SW* 4:391/*GS* 1.2:695). While it is clear what the methodological import of this procedure involves, and while it is premised on and involves different conceptions of interruption and destruction, what endures as the open question is the relationship all of this has to the Absolute, even if it is a reworked, perhaps even reconfigured, conception of the Absolute. Does this other history of destruction entail the effective presence of the Absolute? This is a question that cannot be ignored.

The passage from *The Arcades Project* that opens up the interruption demanded by the caesura and which will allow the question of the Absolute to begin to be posed is the following. It should be remembered that bringing the Absolute into focus is not to add on an extra element. Criticism, as noted already, is unthinkable except in relation to the Absolute. The obvious question is: Does this remain the case given the already noted move from the critic to the historian? Were critics and therefore criticism a form of historical materialism *avant la lettre*?

> Thinking involves both thoughts in motions and thoughts at rest. When thinking reaches a standstill in a constellation saturated with tensions, the dialectical image appears. This image is the caesura in the movement of thought. Its locus is of course not arbitrary. In short, it is to be found wherever the tension between dialectical oppositions is greatest. The dialectical image is, accordingly, the very object constructed in the materialist presentation of history. It is identical with the historical object; it justifies its being blasted out of the continuum of the historical process. (N10a, 3, trans. modified)

The significance of this passage is twofold. Not only does it reinforce the interconnection between interruption and the historical object, but there are also intimations of how a reworked conception of the Absolute can emerge. (These intimations will need to be connected to another passage from the same Konvolut [N2, 3] in which the nature of "historical understanding" is redefined in temporal terms.) Despite being interrelated, each of these moments needs to be treated in turn.

The "dialectical image" is an interruption. As a dialectical image rather than as a simple image, what it involves is the copresence of what can neither be reconciled nor rendered synthetic. The image becomes a type of temporal montage and therefore should not be understood within the conventions of the image. Those conventions will always privilege sight over language. The "tensions" inherent in the image are there precisely because of the impossibility of the image's incorporation into the temporality of historicism or into the procession of concepts and activities that are articulated within that temporal unfolding. This image is described as "the caesura in the movement of thought." What does this mean? Any answer to this question has to begin with the recognition that for Benjamin the dialectical image is the true historical object. Even though that will be a contested assertion, it is the ineliminability of the conflict that directly confirms the impossibility of withdrawing the historical object from questions concerning for whom, and in whose name, a given history is being formulated. Historicism will always try and incorporate "events" into its own conception of continuity. The caesura is the interruption of that attempt. What that interruption demonstrates is that destruction reconfigures both the historical object and what can count as historical. In the same process, it indicates that continuity (whether it be in terms of the naturalization of chronology or the incorporation of myth into and as history) is always a secondary effect whose primary intent is the elimination of conflict—even if that elimination is only ever putative. Chronology, myth, and nature would be terms deployed within the desire for what is always the same. The claim here is that not only does the "caesura" overcome that possibility; it also shows the "always the same" to be a politically charged aspiration and not one that contains the truth of time. In other words, the caesura, in overcoming all that which is entailed by continuity, achieves this end by staging the truth of time. It is precisely this staging that opens up the Absolute.[7]

Truth is not being counterposed to appearance. Even though continuity is an appearance and the truth of time emerges with the interruption of that appearance, there are two additional points that need to be noted.

The first is that the move from *poesis* to *techne* allows for the presentation of time in this way, since reproducibility is already implicated in the reconfiguring of time. Second, the reason why there is no straightforward opposition between truth and appearance is that there is no presentation of truth that has the same status as any given narrative of continuity. There is no narrative of truth. There are only moments of interruption. These moments are fleeting, appearing and disappearing as sites of philosophical and political activity. There will be no final summation. And this lack of finality is not the identification of the Absolute with a domain of unfettered freedom. How then, in this context, will the Absolute figure?

The answer to this question should now be clear. The Absolute is time. Neither chronological time nor clock time, the Absolute is given within the interruption in which the truth of time is presented. Interruption is only possible because what can be known and therefore what functions as the ground of what can be known are not identical with what appears. Knowing what appears, allowing it to be reconfigured as an object of knowledge, necessitates understanding appearance as an effect. There is the inevitability of interruption. It is connected to the way Benjamin defines "historical understanding" as what is "to be viewed primarily as an afterlife [*Nachleben*] of that which has been understood: and so what came to be recognized about works through the analysis of their afterlife, their fame, should be considered the foundation of history itself" (N2, 3). The point being made here is a redefinition of history. Within that redefinition, history becomes the continuity of the reworking of what is already there. This reworking is occasioned by the interruption of the given. With that interruption, what is given comes to be given again and in so doing has an "afterlife." It is, of course, never given again as the "same." This is a process without conclusion. Or at least it is a process whose conclusions are always strategic and provisional. The Absolute is therefore that which allows for the interruption; but equally it is what is evidenced by that interruption. There can be no attempt to present the Absolute, nor even to state the truth of time. The Absolute as time is what allows for the "dialectical image" while precluding any image of time. The absence of the latter is, of course, the moment in which modernity appears as secular.

Interruption as a defining motif in Benjamin's thought dominates both his engagement with Romanticism and his move to the writing of another construction of history. In both instances, the interruption—analyzed in this context in terms of the caesura—is unthinkable outside its relation to the Absolute. In regard to Romanticism, the presence of the Absolute is explicable in terms of a retention of key elements of Schlegel's philosophical and critical project.[8] In the case of *The Arcades Project,* the Absolute returns as time. Two important conclusions can be drawn from this setup. The first can only be noted, while the second can be pursued in greater detail. The first is that it must force a reconsideration of the role of the Absolute within philosophical thinking; even that thinking whose ostensible concern is a theory of modernity. The second is connected, insofar as what must be taken up is the extent to which a theory of modernity will depend upon a philosophy of time that has its point of departure in Early Romanticism rather than in the march of teleological time implicit, for example, in Kant's construal of the relationship between history and the Enlightenment.

Kant is at his most explicit in this regard in a number of his shorter, perhaps more directed, writings. A precise formulation of his position can be found in his 1784 text "Idea for a Universal History with a Cosmopolitan Purpose." What is especially interesting about Kant's formulation is the way it identifies the importance of conflict as a generator of social change. Conflict, however, cannot be taken on its own. It has to be understood as presented within an inherently teleological conception of historical time. The Fourth Proposition of the essay makes this extremely clear. Kant argues that

> by a continued process of enlightenment, a beginning is made towards establishing a way of thinking which can with time transform the primitive natural capacity for moral discrimination into definite practical principles, and thus a pathologically enforced social union is transformed into a moral whole.[9]

The move is not simply from discord to accord. More emphatically, the latter has its conditions of possibility in the former. Contentment as a starting position would, for Kant, trap the human at the position of the animal; the goal of "rational nature" can only be attained by reason coming into play because of the demands made by discord. Even if accord is not identified with a utopian position—and this would be justified, since accord is the result of a passage through time rather than a posited end whose achievement can only be posited—it remains the case that the underlying presupposition is a continuity of resolution within a smooth, hence linear, passage of time. There is no space in which it would be possible to locate the presence of a productive interruption. Linear time cannot allow for a form of recall charging the present with a renewing of intensity. If the caesura stages the truth of time, then while Kant's identification of the role of antagonism and conflict may open up a way of locating a type of discontinuity which may have a productive dimension, what cannot occur is a politics of time. The Enlightenment tradition allows for a way of connecting time and politics, however, since the passage of time is a linear progression—one form of which is historicism—and time itself cannot be sundered in the name of another possibility. Moreover, while the modern can be identified with the centrality of reason and thus with the secular, modernity as an interruption cannot be thought. Equally as important is that the antagonism of these two conceptions of historical time—Benjamin's as opposed to the one implicit in Kant—cannot be accounted for in Kantian terms. Moreover, their presence as possibilities within the modern means that interruption will continue to figure, since the hold of continuity makes modernity an unfinished project.

2. The Time of Fashion: A Commentary on Thesis XIV in Walter Benjamin's "On the Concept of History"

One of the complex problems which endures for any analysis of Walter Benjamin's conception of historical time is the role fashion plays within it.[1] Part of the contention here is that references to fashion in Benjamin's writings work in two related though nonetheless distinct ways. In the first instance, there is the phenomenon of fashion, with its shifts in both dress sense and sensibility. In this regard, Benjamin's concern is with how these changes are registered and with the meaning of that registration. They are not mere epiphenomena. Rather, they are part of the construction of culture itself. This sense of fashion is bound up with the utopian impulse he identifies in the conception of commonality to which the phenomenon of fashion points. (Though it should be noted that this reference to commonality is not simply utopian; it occurs because of the relation, as envisaged by Benjamin, between fashion and advertising; see K2, 6 and K2a, 4 of *The Arcades Project.*) In the second instance, there is fashion as a register of time. In this sense, fashion is inextricably linked to a certain conception of historical time, one that opens up what has been noted on several occasions, namely, the inherently political nature of time. Even though most of Konvolut B of *The Arcades Project* deals with the former sense of fashion, there are a number of important moments where fashion is present as a temporal marker. One of the most significant is B4a, 1. While the majority of the text comprises a quotation from Brecht, it will become clear that time is the actual subject matter.

> The following remark makes it possible to recognize how fashion functions as camouflage for quite specific interests of the ruling class. "Rulers have a great aversion to violent changes. They want everything to stay the same—if possible, for a thousand years. If possible, the moon should stand still and the sun move no farther in its course. Then no one would get hungry anymore and want dinner. And when the rulers have fired their shot, the adversary should no longer be permitted to fire; their own shot should be the last."

Even within the context of *The Arcades Project,* this citation must function in ways to be determined. Never expecting to be reproduced in this form, what can therefore be questioned is the effect of its position in the confines of the presentation of fashion as clothing. Here in this passage, it is *clothing,* though now made specific, clothing as "camouflage." (And hence never just clothing.) This form of clothing will enable something to happen. What occurs, of course, is that fashion—now as a temporal marker—appears as change. The condition of this appearance is that a certain sense of continuity is maintained. The continuity in question is, to use the formulation of both Benjamin and Brecht, undertaken in the name of the "ruling class." This formulation needs to be set productively against another extract from the same Konvolut. This time, in an act of self-citation, Benjamin begins to establish how a connection between time and fashion is to be understood. In B3, 7—a position repeated in greater detail in N3, 2—time is given a more complex register. He writes, "In my formulation: 'The eternal is in any case more the ruffle on a dress than some idea.' Dialectical Image." Clearly, on one level, this is a claim that if not contradicted is certainly mediated by the opening arguments of "The Work of Art in the Age of Its Technological Reproducibility," where the "eternal" is identified among a series of ideas, or if not ideas then concepts or values, that figure significantly in the reception of the work of art. A reception challenged by the advent of reproducibility.[2] In this instance, the differences between them would be slight. As such, what returns is the question of the nature of the distinction between the "idea" and the "ruffle." The question returns because what has to be taken up is the meaning of "ruffle" within such a context. Precisely because the ruffle's presence one year could be filled by its replacement and hence by its absence the next, it can be argued that, whether present or replaced, what is evoked is the same state of affairs.

What then does the ruffle stand for? There is, of course, the more insistent question: How could its very ephemeral and transitory nature possibly be linked to the eternal? These questions will prepare the way for a more sustained engagement with "fashion" as it is present in Thesis XIV of Benjamin's essay "On the Concept of History."

The idea—as opposed to the ruffle—attains eternality on one condition. While obvious, it is worth stating. The condition is that the idea is conceived within idealism. As such, it would acquire the ontological and temporal considerations that idealism confers on the idea. The temporality of idealism cannot be readily disassociated from the eternal. The contrary is also the case; namely, that once divested of such a formulation, the idea occupies a different place. To the extent that there can be a materialist conception of the "idea," or even in stripping the idea of its link to the eternal by construing its presence dialectically, it reappears. The ruffle, however, in being transitory, is engendered by the conception of change generated by an idealist conception of the idea. In other words, with the concession that the ruffle is indeed transitory and that fashion can be understood as the expression of change and movement—all that is initially counterposed to the eternal—there is the related admission that all these conditions are contingent upon the effective retention of the eternal understood as the always the same. This is, of course, the force of Brecht's position. When the quoted passage ends with an evocation of the permanent, and enduring—hence "eternal"—presence of rulers, what is evoked is a conception of time that permits change, though only on the condition that it is located in that which has a rapidity that leaves its ground untouched. Everything must remain in place. While what remains, and thus what stays, will always have content, what is of greater significance—because it allows for that content—is the temporality that such a positioning articulates.

The ruffle opens up the way the eternal figures within the everyday. It figures by isolating change; construing and limiting it to the temporality of fashion. Fashion may bring with it other dimensions. What matters is using fashion against fashion (thereby opening up another possibility for fashion). This can only occur by refusing the reduction of fashion to clothing. Fashion is the site where time endures as overdetermined. And yet, the question of the role of clothing cannot be refused. How are elements of culture—its artifacts—to figure within a concern that identifies fashion with time? In part, the answer to this question is already present in Thesis XIV. When Benjamin argues that fashion has a concern with the "actual present" *(das Aktuelle),* what he has to allow for is fashion's inherently overdetermined nature. Equally, in the confines of Konvolut B, arguments that connect fashion to both death and sexuality all rely on this positioning. For example, in the claim that fashion is "a bitter satire on love" (B9, 1) because it seeks to eroticize the inorganic at the expense of the organic, this once again is an argument that depends upon the complexity of the term "fashion."

The final terms that have to be introduced—almost as a prelude to the detail of Thesis XIV—are "novelty" *(die Neuigkeit)* and "nouveauté." Fashion has an important relation to the world opened by these terms. It is not surprising that they are defined in relation to each other in Konvolut D—a Konvolut whose ostensible subject matter is "Boredom and Eternal Return" ("Langweile und ewige Wiederkehr"). D5a, 5 draws the terms together.

> To grasp the significance of nouveauté it is necessary to go back to novelty in everyday life [*die Neuigkeit in täglichen Leben*]. Why does everyone share [*teilt*] the newest thing with everyone else? Presumably, in order to triumph over the dead. This only when there is nothing really new. [*So nur, wenn es nichts wirklich Neues gibt.*]

The detail of the last line is vital. What is occurring, takes place "only" *(nur)* when there isn't anything that can be described as "really new" *(wirklich Neues).* This "only" creates a scene that moves in two directions. In the first instance, it accounts for why there is a sense of commonality in regard to novelty. It "only" occurs because of the absence of the "really new." In the second, it opens up the question of the "really new." While it may be demanding too much of the passage, it is nonetheless possible to detect that the reference to "sharing"—and therefore to a sense of commonality—is to the utopian impulse that is there, for Benjamin, in the manifest world of the everyday. And yet which is only realizable in breaking through that world. Equally, the reiteration of terms announcing the new, its possibility and impossibility, and thus the question of the "really new" even though only present as a question, means that what has to be taken up is the connection, if real connection there be, between this utopian impulse and the implicit politics of time at work in the question of the "really new." Part of the argument will be that it is only by rendering this conception of the utopian otiose that a politics of time can actually emerge.[3] Behind such an argument is the supposition that the community of the "newest thing"—and this will be the case regardless of how that community is construed—brings with it a conception of commonality that is inherently incompatible with the one implicit in a Benjaminian-oriented politics of time.[4]

Thesis XIV

In its entirety, Thesis XIV of "On the Concept of History" reads as follows:

> History is the subject of a construction [*einer Konstruktion*] whose setting is not homogeneous, empty time, rather a setting filled by the presence of nowtime [*Jetztzeit*]. So for Robespierre ancient Rome was a past charged with nowtime [*Jetztzeit*] which he blasted out of the continuum of history. The French Revolution understood itself as the recurrence [*weidergekehrtes*] of Rome. It cited [*zitiert*] ancient Rome the way fashion [*die Mode*] cites the dress of the past. Fashion has a sense for the actual present [*das Aktuelle*], no matter where it moves in the thickets of the past [*das Einst*]. It is a tiger's leap into the past. Only to find itself instead in an arena in which the ruling class gives the commands. The same leap in the open air of history is the dialectical one, which is how Marx understood the revolution.[5]

Opening a commentary on this passage has to begin with the distinction announced in the opening lines between history as "a construction" *(einer Konstruktion)* and the emptiness of simple chronology. The latter exists as a continual state of ever-moving time in which moments are subsumed within continuity. Their subsumption empties them of content. The continuity of time is therefore mirrored in the homogeneity of the context demanded by such a conception of historical time. The distinction between these two conceptions of history involves the difference between a constructivist conception of the historical—a construction linked to the act, thus the moment, of establishing history through an interruption of simple chronological time—and a conception that is organic in orientation.[6] In regard to the latter, history would be the continuity of its unfolding. For the constructivist position, history does not have to do with time per se, but with a construction arising from an act. That is why for Benjamin history occurs; it is a happening that takes place in a different setting, one "filled with nowtime." "Nowtime" involves the ascription of a potential to time. However, the potential is not for the continuity of its movement. The potential lies in moments of the past. Moments that open up for construction allow for the creation of the historical. The question of how the past is the subject of history misconstrues history—it naturalizes and thus fails to see the historical as the result of a productive act—though equally it misconstrues the event of history. That event is both an interruption and the release of the moment's potential. The release is the construction. However, the site of construction is not given by the passage of time. The site is the present; within it the past is a continual creation. The interplay between potential and the present provides a setting to which it will be important to return.

The relationship between the French Revolution, in terms of its own self-conception, and ancient Rome has to be understood within the terms set by the relationship between potential and the present. Benjamin's formulation needs to be given the interpretive precision that it demands. What matters, in this context, is that the "Revolution" saw itself as "the recurrence [*weidergekehrtes*] of Rome." The significance of the use of these terms lies, initially, in the fact that half of an entire Konvolut—i.e., Konvolut D—is devoted to a critique of "Eternal Return" ("Ewige Wiederkehr"). Indeed, the final note in the Konvolut seeks to establish the complementary relation between the "belief in progress" and the "representation of eternal return" (D10a, 5).[7] For Benjamin, "eternal return" is mythic because it is "urgeschichtlichen" and therefore "it does not reflect" (*es nicht reflektiert;* D10, 3).[8] Reflection is not just a cognitive activity. Far more is involved. Reflection is part of the construction of history. It has to be understood as that which interrupts the hold of myth by allowing it to emerge as myth. What this means is that mythic time never appears as such. The demythologizing move which, in breaking the eternality of the always the same, does so by the construction of a moment. (Hence the opposition between myth and history.) This event, however, does not simply occur. Nor, moreover, is it the result of mere spontaneity. A more complex sense of the moment of interruption has to be given than a mere spontaneous occurrence. The activity of construction will involve what Benjamin identifies as "cunning." (Cunning does, of course, pick up the actative dimension within "reflection.")

> The genuine liberation from an epoch [*Die echte Ablösung von einer Epoche*], that is, has the structure of awakening [*hat die Struktur des Erwachens*] in this respect as well; it is entirely ruled by cunning. Only with cunning, not without it, can we work free of the realm of dream. But there is a false liberation; its sign is violence. (G1, 7)

"Cunning" describes that intervention in which history emerges. Not the making of history but its "construction." The "epoch" is a term used to describe that setting in which the passing of time takes place under the rubric of the always the same. Interrupting that movement—an interruption understood as a "genuine liberation"—does not have a predetermined appearance. Rather, and this is the vital point, it has the "structure of an awakening." Once viewed as a structure, its contents are determined as much by the setting as they are by possibility. The interplay of both is the nature of strategy. For a strategy to be effective—to have a result in the "arena in which the ruling class gives the commands"—it has to deploy cunning.[9] Strategy is all. Insisting on structure allows for history to be a construction. The construction involves a form of repetition that breaks the hold of myth. However, this is not a claim about the content of myth. It is a formal one concerning the temporality of myth. In fact, formalism is the precondition for a politics of interruption. If it were simply a matter of repeating identical content, then the force occasioning "awakening" would have been dissipated. Formalism allows, in the precise sense of making possible, the discontinuous continuity of strategy. Strategy becomes cunning. Cunning, of course, becomes strategy. One could not appear without the other.

In order to explicate further the sense of repetition in Thesis XIV, it is important to note that it is defined by Benjamin in terms of a concept of citation. On a general level, the point being made is that the French Revolution repeated Rome in the same way that fashion cites the "dress of the past." Citation is by its very nature a decontextualization. It is, of course, a decontextualization which concedes the presence of a context. (And yet the gesture that insists on fidelity to context merely repeats historicism's commitment to the smooth passage of time in which the continuity of the always the same predominates.) Fashion is of interest in this regard precisely because it refuses the determining hold of context. The refusal of that hold is evidenced in citation. The citation becomes in the realm of fashion what the "construction" would be in relation to history. And yet, fashion is not just an abstract possibility. Fashion has a "sense" *(Witterung)* for the moment. Another way of identifying this "sense," a way which, while still accurate, brings in an additional element, would be to see it as an occurrence; one bordering yet not being the same as one following from cunning. The similarity between history and fashion—and it is a complex similarity that will in the end have to be questioned—lies in the fact that in both instances a relation between the present and the past is central. A relation established in a moment; thereby establishing the moment. Note that Benjamin does not say history, as though there were, or even could be, an identification of history with that which is past. The sense for the moment that fashion has concerns its capacity to move through the "past" in order to enact a form of repetition that is completely dependent upon decontextualization. The latter is not to be understood as a claim about meaning. It is one about time. Decontextualization is the blast that renders impossible the "continuum of history." The actualization of history, understood as a construction—the emphatic moment—is predicated upon this impossibility.

Again, it is important to note that Benjamin is presenting this position in terms of citation. He is not claiming that the French Revolution saw itself as Rome. It was not an act of empathetic identification. The repetition of Rome, in no longer allowing the context of Roman history to determine the meaning of Rome, allowed Rome to be brought into conjunction with the Revolution such that in their relation of non-relation the identity of the Revolutionary period would have been present in terms of discontinuity. This relation of nonrelation is another way of providing the "dialectical image" with an adequate formulation. What is significant about this conception of the image is twofold. In the first instance, it is not dialectical if that were thought to involve a synthesis. The opposite is the case. It works as an image because of the impossibility of a form of conflation. There could not even be any form of confluence between them. Second, the impossibility of synthesis has to do with temporal concerns rather than with the specific nature of the image. The montage that this setup creates has to be understood as temporal in nature. The juxtaposition of different temporal conditions, what is identified in the language of the thesis as a form of citation, gives the image a formal presence. What it would mean to reiterate this presence cannot have anything directly to do with the specific examples used by Benjamin in *The Arcades Project.* The image cannot be repeated in terms that have similar context, as though that were all that is involved. This is not eternal return. That would leave the image singular both in terms of content and temporality. The image is in fact timed; it involves the copresence—at the present—of that which is temporally discontinuous. This is, of course, the point made by Benjamin in *The Arcades Project* (N7, 7). "In order for part of the past to be touched by the present instant there must be no continuity between them." What matters is this temporal dimension. Hence the question of what it means to be Roman now.

While Benjamin famously describes fashion as the "tiger's leap into the past" *(der Tigersprung ins Vergangene),* this conception of the past has to be understood not as counterposed to the present, but to the "now." Indeed, when describing fashion as having a "sense of the actual present" despite moving in the "past," the "past" is *Einst* which is counterposed to the *Jetzt.* Even though it is the "past" *(Einst)* in contrast to the "now" *(Jetzt),* there is, nonetheless, an associated misconception if that leap is taken as an end in itself.[10] The leap is sited. It occurs in a setting. Indeed, what is important is to contrast its setting to the one identified in the opening line of Thesis XIV. The constructivist conception of history involves a setting that was known in advance. The spontaneity of the "tiger's leap," while an act that can be explicated in terms of an interruption of time, yields a conception of historical time which, even though productive, has a productivity which is importantly delimited. It occurs when it is accompanied by a certain perception of the "arena" of activity. Fashion is oblivious to the nature of that arena. To use Benjamin's own language of dreaming and awakening, it is as though fashion remains asleep even if it enacts—as a possibility, a possibility realized through citation—a form of interruption. It is, of course, not an actual interruption since it is marked, superficially, by a slumbering innocence. On another level it is an innocence that is inherently complicit with the arena where it occurs. What is complicit is not the temporality as such—though whether there is ever historical temporality *as such* remains an open question—but the intricate relation between time and "place" *(Ort)* in which that relation figures. This is the point that has to be pursued, since what is introduced is a complication.

Fashion has the capacity to interrupt. The interruption occurs through the act of citation. Citation is decontextualization and thus recontextualization. And yet, the question that cannot be avoided concerns the extent to which citation on the level of fashion can be taken as the model. It is worth recapitulating at this point. Fashion has the capacity to establish an affinity between the dress or costumes of the past and the present. It does this by allowing for a form of repetition. What marks the realm of fashion out is that such a movement—the movement of repetition enables contexts to exert their hold—is unfettered. What fashion creates is that which comes to be à la mode; its becoming thus will have no real restriction other than the operation of fashion itself. What operates is an industry. Nonetheless, it is an industry in which the process of interruption can figure. Rather than the historical moment, what occurs is its double. However, the interruption does not take place in a neutral setting. While fashion may involve a "tiger's leap," were that leap other than one occasioned by fashion for its own ends, then another state of affairs would have occurred. This is why Benjamin concludes Thesis XIV with an evocation of the "open air of history."

Moments

Is fashion located in a closed domain? This question evokes Benjamin's description of Jugendstil as "the dream in which one is awake" (*er ist das Traümen, man sei erwacht;* K2, 6). This dream is not recognized as such. Freedom appears as the absence of constraints. As an appearance, however, its interruption brings the inescapable hold—under present conditions—of constraint into view. Fashion, or more exactly the modality of fashion, can be understood as involving similar circumstances. Its freedom gestures to the nature of temporal interruption. Its practice involves calculation. Its failure is that it thinks that its field of operation is free. Here is the point at which the question concerning the relationship between potential and the present can be reformulated. In the first instance, the question of potential was posed in relation to history as an act rather than as a description. In other words, history is not the description of the past as though there was a continuity between past and the present. History is an act made in the present and which releases the capacity inherent in the past for it to become historical. History—the act—releases the past's potential to become historical. However, this is not a simple activity. Such acts involve positions—and relations—of power. While fashion only evades a historicist conception of time because its sense of the present allows it to work through the past, in the end, of course, if the "ruffle" can stand for fashion, its very act of evasion becomes the strongest evidence for the retained effective presence of the always the same. Nonetheless, what has to elude fashion is the potential of its historical objects. Fashion operates with a conception of history as the past. Admittedly, it is a past that can be brought into the present via an act of interruption. However, the objects remain of the past. In lacking actuality they lack history. What counts as prehistory, if the argument of Thesis XIV is retained, has to do with a potential that is always present. It is the potential for the emergence of history. The prehistoric or the archaic are present in the precise sense that there is the possibility of interruption. To the extent that the present contains the possibility of being discontinuous with itself, and that this possibility defines the present rather than the future—investigating the future is, after all, a form of blasphemy—then the copresence of the prehistorical and the present defines a setup delimited by potentiality.[11] While this other presence—prehistory—may take on an air hovering on the mystical, such a possibility is there only to the extent that history is not understood as involving an act. Moreover, it has to be an act in which history comes to be constructed. In order that such an act occur and thus the dialectical image appear—the appearance of which is the emphatic presence of a temporal montage—there needs to be activity. And action cannot be separated from cunning.

Here it is essential to turn to one of the most overdetermined formulations of time in "On the Concept of History." When Benjamin writes that "every second of time" *(jede Sekunde)* is itself the gate through which the "Messiah can enter" (*SW* 4:397/*GS* 1.2:704), the "every second" is chronology's measure of time.[12] It is the time expressed within history as a "sequence of events" (*die Abfolge von Begebenheiten; SW* 4:397/*GS* 1.2:704). Rather than defining the present, these "seconds" have to be defined in terms of a potential. The potential is not there, however, in a yet to be determined future that is always to come. It is there in the possibility of an interruption, which is the occurrence of history. As such, the present as the passage of "seconds," the past as events awaiting history, define a moment that is continually charged. Robespierre's understanding of ancient Rome historicized the present by giving actuality to Rome. As an event it is unrepeatable. As a formal description, it becomes the structure of a form of recurrence that undoes eternal return in the name of the present.

While "every second" marks potential, and while interruption occurs strategically, it is equally the case that the conflict between chronology, which holds both mythology and progress in place, and the interruption which is effected, if not understood, by the modality of fashion, is of fundamental importance to any analysis of the differing temporalities of historical time. Benjamin's expression at the end of Thesis XIV of the "open air of history" demands more than the simple identification of historical consciousness. The contrast between the "tiger's leap" of fashion and the "same leap" *(derselbe Sprung)* as bound up with Marx's conception of "revolution" is a difference that can be explained in terms of what the opening lines described as the "site" *(Ort)* of history. And yet, it is not the site of history as though that site were merely given. It becomes it. Hence, the distinction between the "tiger's leap" and this other leap which, despite being the "same," is nonetheless also fundamentally different. The sameness is explicable in terms of interruption, the difference in terms of place. Again the difference is not straightforward. Time and place are connected. One is positioned by the other. This positioning has important implications in regard to the concept of collectivity that is envisaged by the place of fashion as opposed to the place of history—history as a construction occurring in and defining the present. Pursuing this final point will allow for the preceding threads to be drawn together in the guise of a conclusion.

The utopian impulse that brings a sense of commonality with it, and which is directly connected to fashion, is presented in the following terms:

> Fashion, like architecture, inheres in the gloom of the lived moment [*in Dunkel des gelebten Augenblicks*], belongs to the dream consciousness of the collective [*zähen zum Traumbewußtein des Kollektivs*]. It awakes, for example, in advertising. [*Es erwacht—z.b. in der Reklame.*] (K2a, 4)

In this context, what is significant about these lines is that they identify fashion as having a link to an archaic sense of the common. The dream consciousness becomes the utopian possibility. The possibility finds actuality in the community of "advertising," the latter being the sense of community envisaged by the advertisement. However, it is the only sense in which the advertisement is intended to appeal to all. Again, this brings with it the implicit commonality of the department store, or even the arcade. And yet, while there may be a sense of commonality it is not the commonality of the "clear air of history." It is the commonality that takes "place" to be unproblematic. It is not just that it is the commonality of fashion; rather, the important point is that it is a sense of commonality that cannot occur in the "place" of history. It is incompatible with that place. And if time and place are interarticulated, then it is antithetical to the conception of historical time bound up with this sense of place. What is recalled, therefore, is the community of the "newest thing."

If there is an overall consistency in Benjamin's position, then it has to be that the "tiger's leap" mimes the possibility of history. Fashion works as "camouflage." The "ruffle" brings the eternal into play since it presents a possibility of change, and thus the new, that is predicated upon the retention of the always the same. Nonetheless, fashion counters itself. It has a "sense" for actuality. However, it is only a sense; a nose for what is. Fashion involves a form of "cunning" to the extent that its status as camouflage is maintained. What, then, of the community of the newest thing? How, then, is the collectivity promised by advertising to be understood? (A collectivity gestured at by fashion.) Answering these questions necessitates insisting upon the distinction between the temporality of history and that of fashion. Fashion mimes history. The reason for using the word "mime" is that fashion cannot create history. If history is a construction, one in which the past occurs in the present—and can only occur as such—then while fashion allows the past to appear, it only ever appears as the past of historicism, not the past of history. History is given by the interplay of construction, time, and place, one in which place is the site of recurrent contestation. The utopian gesture that refuses place cannot, as a consequence, think the particularity of historical time. In addition, time as given by the unfettered leap cannot have an adequate conception of place. While this may be taken as the force of the utopian, it also indicates why utopianism is more than just an empty gesture; it is one based on a profound misconception of time and place.

The only way that a move between the utopian as a possibility and its realization could occur is that the utopian contain a potential. This is a possibility that has to be present even in the gesture that defines utopianism. Potentiality would have to be present in order to save utopianism from becoming a politics deprived of either goals or ends. This is the point to introduce the epigram, taken from a work by Karl Kraus, and with which Thesis XIV can be said to begin. The four-word citation is straightforward: "Ursprung ist das Ziel" (Origin is the goal). Leaving aside its actual source in Kraus—since its decontextualization brings with it a framing of the complex figure of the "leap" *(Sprung)* in Thesis XIV—it would be productive to read the recourse to an origin *(Ursprung)* not as an initial point of departure but as an originating leap. Thus the citation announces this "leap" *(ur-sprung)* as the goal *(Ziel).* If there were a slogan proper to the distinction between fashion and politics—a slogan locating and delimiting the so-called utopian impulse—then it would be that the "goal" be reached. Reaching involves not a simple leap but a movement through the conflict that bears the name politics. The "goal" is neither the utopian moment nor its realization. On the contrary, the "goal" is the open possibility. An opening occasioned philosophically by potentiality, though realized, in the "open air," politically. What this means is that despite a real equivocation within Benjamin's writings concerning utopianism, what arises from an engagement with the politics of time at work in Thesis XIV is that political possibilities are connected to potentiality. Potentiality, in this context, results in the construction of history. A construction realized through strategic means; means that can always be recast in terms of cunning. Noting, of course, that neither means nor cunning is driven by a goal or end if either term is futural. The interruption—figuring, for example, in the dialectical image—has to be understood as an allowing. Moreover, it is moment that transforms the temporality of the "every second." The transformation becomes possible because the past contains a potential. The form of that realization cannot be predicted. Moreover, it cannot have an already determined content. This is why Benjamin argues, as has been noted, that a "genuine liberation from an epoch" has the "structure of an awakening." Awakening is the construction that is the allowing. The formalism provides for history and marks the force of strategy. In contradistinction to this positioning, utopianism empties time by giving the future an already determined, even if idealized, content. Utopianism cannot sustain potential. Within it structure and content have to be given in advance. Such a possibility is predicated upon the effacing of potential. Only through the retention of potentiality and a formalism that will always allow for content—a possibility actualized by the moment of interruption—will a politics of time be possible. What is at stake is an interruption; its possibility depends upon potentiality; its occurrence allows. Rather than the cessation inaugurated by violence, here the interruption—what for Hölderlin would have been the countermeasure of the caesura—is occasioned strategically. After all, "only with cunning, not without it, can we work free of the realm of dream" (*AP* 907/*GS* 5.2:1213).

3. Benjamin's Style: The Style That Is Not Jugendstil

With Jugendstil, the question of style returns. It is not as though, in the German context, the question had not been raised before and had not given rise to an extended architectural, artistic, and political debate.[1] (It should be noted that this is a question that dominates the intellectual agenda in these areas and it will return, perhaps in the wake of Jugendstil, with Walter Curt Behrendt's 1927 work *The Victory of the New Building Style.*)[2] The question's inception, in the realm of architecture—an inception in which forms of materialism came to be counter-positioned against forms of idealism—was the publication in 1828 of Heinrich Hübsch's pamphlet *In What Style Ought We to Build?*[3] Prompting this latter work was a climate in which the relation to antiquity had emerged and continued to emerge as a fundamental concern. That emergence, however, has to be understood as staging a more exacting question. And here it is essential that this staging be recognized. Any argument concerning antiquity needs to be read as a symptom. What is at work within arguments of this type is not antiquity as such. In the same way, the debate concerning polychromatic antiquities which was sparked by Jacques Ignace Hittorff with the publication of his *Ancient Architecture of Sicily* (1827)—and which exercised a generation of architects, including Gottfried Semper—was as much a concern with archaeology as it was with the nature of European modernity, and indeed of European identity itself.[4] Antiquity was the construct in relation to which the nature of the modern was taken to be at work. What will be argued in this context is that the work of Walter Benjamin, and in particular both the implicit and explicit rethinking of historical time in *The Arcades Project,* makes precisely this conception of the modern problematic. The advent of modernity, understood as an interruption, will consist of more than Europe becoming troubled as to its own identity. And yet it should not be thought that as a self-identified question, modernity is not in some sense present. Time and history figure within a concern with the present's relation to antiquity since, as has been intimated, what is at stake in the question remains a conception of modernity that is thought as a dislocation demanding action. Whether that action is the recovery of a connection or the affirmation of its impossibility—a recovery or an affirmation that will be uniquely concerned with appearance—is not the point. There is a sense of the modern, and even a thinking of modernity, within such a setup. Part of what marks it out is the identification of the modern with the positing of either its presence or its need. In order to identify such a conception, and more important, to distinguish it from that thinking of interruption that characterizes Benjamin's work, and to be consistent with the argument developed thus far, it will be designated as the modern without "cunning." In sum, this is the positing of the modern. As though affirmation were all that was involved.

Not surprisingly, the nature of the relationship to the dominant elements of tradition emerges as sites of contestation precisely at the time in which it is becoming increasingly less possible to maintain any real connection. The question of relation—its ineliminable presence as the question determining any thinking of the modern—while superficially inspired by the current state of archaeological research, or the limit of classical scholarship, is more significantly the question whose emergence is the mark of modernity's inaugurating presence. Once the question of relationality can be asked, then, regardless of the answer, the question announces the possibility of another beginning. Even if that beginning is the resurrection of continuity, and it will always be another continuity precisely because it is the continuity that has to be reestablished as the answer to the question of continuity, what endures is the presence of a beginning. The challenge presented by Benjamin's conception of historical time is how that beginning is to be thought. There cannot be a simple positing of beginnings. As though, with such an act, there will have been, in fact, another beginning. Announced here, therefore, in the reworking of the problem of beginnings, is the distancing of the temporality of fashion.

On Style

In a text of 1818—"Ancient and Modern"—Goethe asserts in relation to the contested heritage of Greece that "everyone should be Greek in his own manner. But he should be Greek." *(Jeder sei auf seine Art ein Grieche! Aber er sei's.)*[5] While this nuances the position of Winckelmann, whose work is central to any understanding of the historicity of style, it is a position that is far more equivocal then the emphatic question posed—as has already been noted—by Heinrich Hübsch only ten tears later. It should be recalled that Hübsch's pamphlet, which inaugurated the debate on style, was entitled *In What Style Ought We to Build? (In welchem Style sollen wir bauen?).* Even though Hübsch will provide an answer to his question with the introduction—perhaps reintroduction—of Rundbogenstil, it remains the case that the decisive elements in the title are the demands stemming from the interplay of the word *sollen* and the question mark. What, after all, is the relationship between a demand, almost an obligation, and its incorporation into a question? The straightforward position is that the answer is to be determined. In other words, Hübsch's question marks the severance of style and appearance. For Winckelmann the connection was given. There was no space within which the question of style could have been posed. Nothing was to be determined. What is at stake in regard to Winckelmann and Hübsch is the interplay between two different conceptions of historical time and the differing roles that mimesis plays within them. In this context, mimesis has to be understood as involving a theory of imitation that is inextricably bound up with the question of historical time. Given, first, that the difference between Winckelmann and Hübsch can be understood in terms set by the relationship between style and appearance—its severance in the case of the latter and their confluence in the former—and second, that time and the process of imitation are connected, it then follows that mimesis plays a central role if the presence of the modern is thought in terms of a founding interruption. (The language of imitation, as will be noted, provides the actual terminology through which they formulate their differing positions.) Winckelmann and Hübsch both play a pivotal role in developing an understanding of the relationship between style and modernity, since their work and the differences between them cannot be thought outside the confines of a specific sense of historical time.

In light of this setting, it is possible to turn to Winckelmann's pamphlet *On the Imitation of the Paintings and Sculpture of the Greeks.*[6] Rather than comment on this work in its entirety, two elements will be of concern. The first is the use of the term "imitation," and the second is the role played by Raphael in his overall analysis. While this may seem a preoccupation that has too great a distance from the question of the appearance of the modern, and in particular modern architecture (that being the question posed by the title of Hübsch's pamphlet), part of what will be argued here is that it is in the move from Winckelmann to Hübsch that the framework emerges in which it will become possible to pose this question. More significantly, as will be argued, it means that it can be posed in a way that the answer remains open.

It is essential to recognize that the key term is "imitation."[7] In his essay Winckelmann writes, "There is but one way for the moderns to become great and perhaps unequalled, I mean by imitating the ancients [*die Nachahmung der Alten*]" (*R* 2/*KS* 60). While this is the favored strategy, it is vital to be clear concerning what is meant by imitation. The answer to this question will emerge from a brief consideration of the role of Raphael. Winckelmann quotes Raphael writing to Baldassare Castiglione in regard to his *Galatea* that, "Beauty being so seldom found among the fair, I avail myself of a certain ideal image [*Einbildung*]" (*R* 12/*KS* 68). What is interesting is that the Platonic problem of moving from the ideal to its realization is absent from Winckelmann's formulation. This absence opens up the question of how imitation is to be understood.[8] He goes on to identify the presence of the ideal in, for example, the "profile of the brow and the noses of the gods and goddesses [being] almost a straight line" (*R* 13/*KS* 68). What is imitated is line and, more generally—though equally specifically—contour. Contour as the object and source of imitation will always stand above nature. Contour allows the move from the individual element to the whole within the practice of the particular work. This, for Winckelmann, was the source of Raphael's greatness, though equally it was the limit of Michelangelo's.[9]

The movement in which "contour" *(Kontur)* and "expression" *(Ausdruck)* come to be connected involves a reinscription of the ideal. Rather than the ideal being a transcendent quality and as such unattainable and unpresentable, it is located in a sense of the appropriate. Note the way Winckelmann plots the effect of "parenthyrsos," a term from Longinus used to plot excess in relation to character.[10] Excess becomes that which is inappropriate. The possibility of excess will lead to arguments for decorum as that which will preclude all of the detrimental effects that result from excess. The example is the ancient Greek sculpture of *Laöcoon.*

> In Laöcoon suffering alone had been "parenthyrsos"; the artists therefore, in order to reconcile the significative and ennobling qualities of his

> soul, put him into a posture, allowing for the sufferings that were necessary, the next to a state of tranquility; a tranquility however that is characteristical; the soul will be herself—this individual—not the soul of mankind: sedate, but active, calm but not indifferent or drowsy. (*R* 32/*KS* 82–83)

Propriety, understood as the appropriate, pertains to that which is proper to character rather than to a transcendental ideal form. Ideals for Winckelmann can be instantiated.[11] This achievement marks the work of Raphael. This opens up the complexity within the proposition that greatness in art is found in imitating the Greeks. What is imitated has to do with a relationship between soul and line, a relationship that, even if the colors of the original fade, the enduring presence of the soul remains. Contemporary practice therefore, in order to work within a Winckelmannian framework, should operate in relation to a predetermined space of evaluation. Winckelmann is able to judge, and the ground of judgment is a founding sense of propriety that has to be imitated. Precisely because it is imitation rather than simple copying, it is possible to see a line between the Greeks and Raphael. However, it is a line determined by continuity. What this means is that while there may be unexpected forms that will allow contour expression—or at least that is a potential whose elimination is impossible—the drawing of a line must be continually guided by this founding sense of the appropriate. This is not a nostalgia for the Greeks in any simple sense. Nor is it a claim that Raphael is Greek. It is an argument for the continuance of a conception of style that regulates appearance. The criteria of regulation are given in advance. While appearance may not be determined in advance—hence the distinction between imitation and simple copying—the question of style is always resolved. The resolution is given within the structure of mimesis. As a structure, it establishes a relationship between line and color (in general appearance) and the ideal. The ideal is always given within the work and articulated—at least in this pamphlet—in terms of the relationship between contour and expression. The other defining element of the structure is time. The question of style is resolved through the activity of imitation; an activity that is itself only possible because of a conception of historical time thought in terms of continuity. Were that continuity not to pertain—and this would mark either a specific moment or a specific work—there would still not be a sustained loss. The continuity can be recovered. Other, possibly futural works, will enact this possibility. Such acts of recovery would be further evidence for why the question of style will always have been determined in advance. In sum, imitation, in the sense in which it is used by Winckelmann, is as much a temporal category as it is one linked to the production of works. Appearance will always be regulated by the interplay of time and imitation.

This is the point at which Hübsch's pamphlet needs to be reintroduced. While Hübsch's own relation to the Navrenne School should be noted—since a purely historical contextualization would locate him as part of that movement—the pamphlet escapes this historical reduction.[12] The pamphlet could not open with a more emphatic diagnosis of the current condition of creation. His language repeats Winckelmann's, since the reiteration of "die Nachahmung" (imitation) joins the texts while at the same time giving rise, almost of necessity, to critical engagement. Nonetheless, what is presented resists equivocation. Instead of arguing for an imitation of the Greeks, the position is the contrary. What has to be explained is why this counterclaim is not just a simple negation of the Winckelmannian position. The pamphlet begins with the following formulation, which amounts to an attack on imitation that is linked to a particular formulation of the modern.

> Painting and sculpture have long since abandoned the lifeless imitation of antiquity. Architecture has yet to come of age and continues to imitate the antique style [*den antiken Styl nachzuahmen*].[13]

What is significant about this formulation is the positioning of architecture as outside the modern—"it has yet to come of age"—and by that position being defined in relation to style. Once identified in these terms, then the range of possible responses is set up in advance. Perhaps the most significant response, and it is one touched on by Benjamin in "Paris, Capital of the Nineteenth Century: Exposé (of 1939)" (*AP* 15/*GS* 5.1:62), is found in the writings of Karl Boetticher. What is important about Boetticher's approach is that he identifies an element that allows the Greek style to have a force that is not defined in relation to the representational quality of the object. In other words, the Greek (or Hellenic) style is dependent upon its material presence. The object is defined in terms of its material possibility. The move from the Greek to the Gothic, and then to the modern, becomes the move from earlier material forms to iron. In terms of material, modernity necessitates iron. The question is the extent to which the introduction of iron will have an effect on the actual style of architecture. For Boetticher the question is a complex one. In part, the response is to be found in the manner of his defense of Schinkel. Within that evaluation, the relationship between the Hellenic and the contemporary is articulated in terms of the centrality of iron. He formulates his position in the following terms:

> The relative strength of iron beams, replacing those made of stone in the trabeated system, can only play an indirect and minor role; to replace the Hellenic trabeated stone system with a trabeated iron system would represent only a change of material, not a change of principle. It would lead to one-sided and very limited progress and would prove as inadequate as stone beams for spanning wide spaces . . . The structural principle is thus to be adopted from the arcuated system and transformed into a hitherto unknown system; for *the art-form of the new system* [*der Kunstformen des neuen Systemes*], on the other hand, the formative principle of the Hellenic style must be adopted in order to give artistic expression to the structural forces within the parts, their correlation and the spatial concept. This alone will create the true mediation, the right synthesis of the two preceding styles. In what manner and by what art forms the structural and spatial manner might be expressed within this newly formed system is a question that the thoughtful person will not find too difficult to answer. [emphasis mine][14]

The answer to this question is found in the work of Karl Friedrich Schinkel. Without pursuing the detail of his interpretation of Schinkel, it is essential to recognize that Boetticher is after a formulation of a sense of continuity, a ground of judgment and the possibility of the new. What is of interest is that the new cannot be disassociated from what new materials make possible. That the new is also connected to fundamental principles or even to the grounding of a style in the nation-state has to be understood in terms of the relationship between external forces and materials that are positioned outside any straightforward structure of imitation. When Boetticher suggests that the form taken by the adoption of the "formative principle of Hellenic style" will be able to be worked out by the "thoughtful person," this leaves open the possibility that such a style is determined in advance, and that the form is "Schinkel"; or, on the contrary, that the style is not determined in advance, precisely because there is no one style appropriate to the incorporation of the Hellenic, where the Hellenic is understood not as a literal style (i.e., thought within the identity of style and appearance), but as the relationship between function *(Zweck)* and material. What is opened up here is a distinction of fundamental importance. On the one hand, there is the identification of the new with a particular determined style. Identification would become a reidentification, a movement in which material carried the determining role. The other is that the new, while inextricably linked to material possibilities, is not given with (or within) a single appearance, let alone one whose determinations are reidentified.[15] It will be vital to return to this distinction, since it is the setting within which the question of style—a question that will continue to encounter the problem of the literal—will have to be pursued.

Staying with the context provided by the text, this problematic status of style needs to be articulated within the division established by Boetticher between the *Werkform* and the *Kunstform* of the architectural object, a distinction introduced by Boetticher in his *Die Tektonik der Hellenen.*[16] As indeed the passage cited above also makes clear in its evocation of the "art-form of the new system." The complicating factor is always going to be how "art-form" is to be understood. This is as much an interpretive question as it is one that involves recourse to perception and thus to appearance. While the *Werkform* allows for the development of materiality as that which is fundamental to the development of architecture, *Kunstform* has only a semiautonomy precisely because it opens up within the public realm, and therefore will have attributed to it the quality of an interpretive object.[17] As such, *Kunstform* cannot be completely extracted from the realm of either symbol or allegory. Once *Kunstform* acquires this status—and it does so precisely because of its semiautonomous presence—it becomes the regulative element within the work. As will become clear, the distinction between *Werkform* and *Kunstform* has to be situated in relation to a more general distinction that needs to be drawn between style and appearance. (In other words, it needs to be invested with another history.)

Returning to Hübsch, it is important to note the force with which he begins his pamphlet. When he writes that architecture should not continue "to imitate the antique style," the suggestion is clear. For Hübsch, the new style will have a complicated relation to any historical precedent. That claim is, however, one made about the appearance of the building. The argument to do with imitation—as formulated in the above—is concerned with appearing. The difficulty with this position is that while there seems to be evidence for a move away from imitation, this is not a straightforward position, insofar as it becomes part of an argument about the relation to history. This is not to suggest that any sense of *Kunstform* is absent. The contrary is the case. However, what is present is the naturalization of *Kunstform.* In connection to a type of populism—one that is played out elsewhere in terms of the relationship between style and national identity—Hübsch will still argue that "the new style . . . will have a truly natural character." What this means is that instead of an allegory that involves permanence articulated through the forms of decoration, there is a recognition of the natural as the popular and thus as a type of natural attitude. Once again, this move to the natural has to be understood as occurring within the setting provided by the distinction between *Werkform* and *Kunstform*. The latter is present as it inheres in the reiteration of an opposition between the material presence of the work and that work having "a truly natural character." What has to be incorporated at this point is the confluence of the "natural"—the form of nature—and the national; one will become the other in that both involve forms of organic unity. (It should be added, of course, that such a unity has a necessarily mythic status.) It follows that what will interrupt the complex relationship between the natural, the national, and the question of style will have to be more than its putative other, i.e., the international. The response to the running together of the natural and the national has to begin with the recognition that what is at work within it is the construction of myth. As a response, and the response has to involve a necessary demythologizing move, it will take place in terms of the *cosmopolitan*. While awaiting greater clarification, this term can be provisionally understood as that which, in refusing the determination of either the national or the international, will demand a rethinking of any politics that involves an inherent commitment to the symbol; even if the latter is inextricably bound up with the process of allegorization through its simple ornamentation.

There is an interpretation of this distinction that holds *Kunstform* back from any type of allegorization, a possibility that is already there in Boetticher. The argumentation would stem from Boetticher's claim that "the structural member and its art-form are initially conceived as a single whole." While this opens a reformulation in terms of autonomy, that autonomy needs to be set against the argument that architecture has to be mimetically related—and that relation will also be organized on the level of the symbol—to the order of nature. Autonomy is always under the sway of nature. The purposiveness of nature understood as the generation of natural form is mimetically related to a possible purposiveness within architectural form. Formulated in this way, however, what is immediately set in play is a relation between form and agency. The aesthetic would become the recognition of the "natural" order of architecture. Once the question of who would be the subject of this position is asked, then what emerges is the argument that has already been noted in Hübsch. The mimetic relation between architecture and nature brings with it a conception of the aesthetic and therefore of agency and subjectivity as the purely neutral and therefore as "natural character."

While it involves a shift of eighty years, it is not difficult to see the presence of the same structure of thought in the lecture "Art and Technique" ("Kunst und Technik") given by Peter Behrens in 1910. In an argument concerning the nature of style, Behrens formulates his position such that *Kunstform* becomes style, insofar as both concern appearance (where the latter is understood as the being present of a work).

> By style we mean nothing but the unified formal expression [*den einheitlichen Formausdruck*], the manifestation of the entire spiritual life of an epoch [*den die gesamten Geistäußerungen einer Epoche ergaben*]. Unified character, not the particular or the peculiar, is the decisive factor.[18]

What is significant in this formulation, and it repeats the point that has already been made about agency and subjectivity, is that it gives rise to the question: For whom is this appearance a style? (More generally, the question is: How does appearance become style?) These questions should not be thought as concerning mere style. Style, formulated and argued for in ways that are evident in both Hübsch and Behrens, is inextricably linked to unity. (Here again, the confluence of the natural and the national.) However, that is only one half of the link, since there is a necessary reciprocity involved. What has to be argued in addition is that style unifies. Once this shift is allowed, and it is a shift that is made possible by the inherent logic within style, then what becomes important is that style cannot be positioned beyond its capacity to generate and sustain unity. Here, in sum, from within an idealist framework is the problem of style. As will be seen, the question of unity—both in a projected but equally in a historic and thus utopian way—will also play a decisive role in Benjamin's argument concerning Jugendstil in particular and the emergence of the commodity more generally.

The reason it is possible to reformulate these considerations in terms of the "problem of style" is that such a formulation indicates the way in which style—thus construed—is presented within what could be described as relations of dependence. Style is articulated and sustained by two relations. The first concerns historical time. Despite the differences at work in the considerations of style noted above, each one brings with it a particular relation to—and more emphatically is the articulation of—a specific conception of historical time. The second is that fundamental to style—arguably what makes it a style rather than an idiosyncratic expression—is that in regard to its aesthetic dimension it is unified in appearance and thus has the effect of unifying. What is unified—a people, a race, etc.—is then able to be identified with a specific style. What these two relations entail is that any critique of style—either as a given style or the conception of style as it comes from this heritage (and in both cases there cannot be any easy separation of style and appearance)—is a critique of the interconnection between style, time, and agency. The latter being the subject position demanded by the unity of a style that unifies both itself and a generalized subject. Benjamin's critique of style, or more accurately, the elements of that critique that have actual force, are implicated in precisely this set of interrelations.

Benjamin and Jugendstil

In sum, the position in relation to which Benjamin's engagement with Jugendstil has to be situated concerns precisely the complex relation between style, unity, and time. At work in the formulations that are linked to the debate around style is a self-avowed conception of break and thus of crisis. This means, of course, that they harbor a conception of the modern. What has been suggested, however, is that the complexity of the situation—a complexity given explicitly, though given with just as much force implicitly, in the retained distinction between *Werkform* and *Kunstform*—is a conception of historical time that allows the work's *Kunstform* to endure as the unified and as the unifying. This will be the case even if the continual transformation of the *Kunstform,* in terms of the materiality of architecture, is to be allowed. This is not to argue that there is not a conception of modernity with the debate—as clearly there is. What is being suggested is that it endures beyond the hold of any activity that inaugurates the interruption. Another way of making this point would be to say that precisely because this conception of modernity reiterates the position in which the new is simply posited, the modern endures without "cunning." Historicism, therefore, becomes a way of holding to a conception of the new without "cunning."

Jugendstil is also concerned with discontinuity. As such, its preoccupation with the modern intrudes into more than matters concerning the particularity of form. Henry van de Velde in *Déblaiement d'art* (1894) writes of a regeneration of art occurring as the result of "the annihilation of the actual" *(l'aneantissement de l'actuel).*[19] Benjamin had already identified this evocation of destruction—of even a productive destruction—as one of the motifs of Jugendstil. In a note of fundamental importance in Konvolut G, Benjamin positions the strengths and weaknesses of Jugendstil within an opposition between "cunning" and "violence"; an opposition set in place by the structure of awakening. The passage, which has already been noted, is as follows:[20]

> The genuine liberation from an epoch [*Die echte Ablösung von einer Epoche*], that is, has the structure of awakening [*hat die Struktur des Erwachens*] in this respect as well; it is entirely ruled by cunning. Only with cunning, not without it, can we work free of the realm of dream. But there is a false liberation; its sign is violence. From the beginning, it condemned Jugendstil to failure. (G1, 7)

While displaying a familiarity with certain of his writings and projects, there is no suggestion that Benjamin has the position of van de Velde in mind when he makes such a comment. What is important here is to see in what way this formulation complicates questions of interruption and discontinuity. Part of what is being suggested in this passage is that the evocation of a "liberation," and thus of a beginning, cannot on its own count as a "liberation." More is involved. And without at this stage detailing its force, it will lie in the use made by Benjamin of the formulation "the structure of awakening." The consequence of its use is that what it sets in play is a position that allows for a distinction to be drawn between "awakening" as a formal possibility and the literal instances that are taken, rightly or wrongly, to enact it. In other words, what such formalism will allow is a way around the positing of the new.

"Genuine liberation," while strategic, is therefore necessarily linked to the "structure of awakening." If there is a possibility that can be generalized, then it inheres in the structure, not in the specific examples given in the body of *The Arcades Project.* It is vital, therefore, not to allow subsequent work to have a literal and therefore nostalgic relation to the content of *The Arcades Project.* Allowing the "structure" a force opens up a form of generality in which the structure is given actuality. Restricting it to the examples given in the folders comprising *The Arcades Project* is to historicize the argument and thus freeze it within the temporality of historicism. Cunning provides a way of thinking the departure from a preoccupation with the literal example. (The study of culture is a study of culture, not its determinations as frozen at a given moment.) Finally, emphasizing structure rather than simple particularity means that what has to be established is a connection between the structure and a possible presentation and thus appearance. What this means in broader terms is that the distinction between "structure" and appearance reiterates the severance of style and appearance that has already been noted in Hübsch. This severance marks the real difference between Hübsch and Winckelmann, and is then mediated throughout the debate on style.

In Konvolut S, the position involving the easy celebration of the new, even a conception of the new that comes to be articulated in the language of "crisis," is diagnosed in terms of the plurality of possibilities held in the term "modern." Benjamin's formulation of this position demands close attention. He argues that "Jede Zeit erscheint sich ausweglos neuzeitig." In other words, "Each time period appears to itself, unavoidably, as a new time." After which he adds that "(The 'modern' however is as varied in its points of reference [*in dem Sinne*] as the different aspects of one and the same kaleidoscope"). Here in the last line is the "modern" as the always the same. It is, of course, not the always identical, but the always the same. Enumerating all of the moments that are given over as modern is to list what he describes as the "traits" of the modern and would be to "represent hell" (S1, 5).

The language of appearance—note that Benjamin's formulation is in terms of the epoch as it "appears to itself"—has to be given a double register. In the first instance, there is the detail of this appearance. Two questions arise. What is it that appears? How is its appearance to be understood? The next element is that appearance allows for its own interruption. Not an interruption that reveals the truth of appearance, as though all that is involved is the truth that lurks beneath or to which it is counterposed. What interrupts appearance and interrupts it in the name of truth—even though the truth in question is a truth that is posed beyond the hold of the opposition between appearance and reality—has to do with the truth of time. In other words, the language of appearance in this context has to do with temporality and not to do with the content of the appearance per se. This will have a dramatic effect on the question of style. If modernity has a style—not the "modern," the term usually identified by Benjamin as always a citation of a particular conflation of the modern and the new, but modernity as a specific thinking of time—then it cannot emerge simply on the level of appearance. It cannot just appear. Without at this stage wanting to address the question, what has to be noted is the problem of the appearance of modernity. What would its style be? (The force of this demand—and it is one to which a response is yet to be given—establishes the ground of judgment.) The argument here is that the advent of the modern is inextricably bound up with the separation of style and appearance—traced in the opening of this essay in the move from Winckelmann to Hübsch—and appearance, as a consequence, becomes a question. It is neither given nor can it be unproblematically posited. Moreover, once the modern understood as the simple positing of the new can be separated from an understanding of modernity as given in the distinction between the temporality of fashion and a more sustained sense of interruptive difference, then the question of appearance becomes all the more complex. Part of what it will entail is the ineliminability of contestation concerning the appearance of the modern, precisely because this will be part of what constitutes modernity.

In order to pursue this newly posed question of style, two of the elements that dominate the positioning of Jugendstil within *The Arcades Project* will be taken up. The first is the relationship between Jugendstil and technology (S8a, 1; S8a, 7; S9a, 4), and the second is the relation to the individual and to individualism; perhaps best articulated in relation to Ibsen's *The Master Builder.* It should already be clear that the reference to the individual and the idiosyncratic has to be understood as the individualizing of style. In other words, it is to turn the concept of style against itself by making that which is universal—or at least should be universal—into something particular. It should not be thought that these two elements are unrelated. In, for example, the "Paris, Capital of the Nineteenth Century" exposé, Benjamin writes:

> The new elements of iron construction—girder forms—preoccupy Jugendstil. In ornament, it endeavours to win back these forms for art. Concrete presents it with new possibilities for plastic creation in architecture. Around this time, the gravitational centre for living space shifts to the office. The real centre makes its place in the home. The consequences of Jugendstil are depicted in Ibsen's *The Master Builder:* the attempt by the individual, on the strength of his inwardness, to vie with technology leads to his downfall. (*AP* 9)

In a passage of equal significance, one that joins these elements together, Benjamin writes that Jugendstil has a positive element insofar as within it the bourgeoisie gains access to the technological and therefore gains "control over nature" (S9, 4). Nonetheless, it also announces a regression, since within it what emerges as impossible is the capacity to look "the everyday in the face" (S9a, 4). The bourgeoisie—knowing its days are numbered—therefore desires youth, longer life, or "death in beauty" (S9a, 4). These two motifs reiterate themes central to Benjamin's overall project. The first is that it allows Jugendstil a place within arguments having to do with the modern. The second is that the evocation of the face and the look recalls the construal of aura provided in the earlier work on Baudelaire in which the aura, or rather the experience of aura, was described in terms of the object—be it a face or a natural object—to look back (*SW* 4:338/ *GS* 1.2:647). The loss of the look is the decline of aura. Here there is the impoverishment of experience; a position emphatically established in Benjamin's 1933 text "Experience and Poverty." With Jugendstil this isolation becomes, on the one hand, the isolated individual—this is, of course, the position of Master Solness in *The Master Builder,* for whom time and the individual come to be interarticulated as a response to the advent of technology and the intrusive presence of both youth and desire—while, on the other hand, style turns from the semiautonomous symbol that works through acts of unification, to the isolated motif that incorporates a literal presentation of nature. Instead of allowing for a mimetic relation in which architecture acts out nature's purposiveness, it literalizes nature, turning it into a cache of symbols. Mimesis collapses and imitation—a path from the idiosyncratic that can only ever turn back on itself in increasingly more stylized circles—emerges in its wake. The move from mimesis to imitation is of fundamental importance. What is occurring is a move away from a relation that is inherently formal—formal insofar as it involves a relation between the purposiveness of nature and the purposiveness of the architectural object—to one in which style becomes identified with a literal style. However, precisely because that literal style is singular and thus idiosyncratic, it cannot fulfill the condition of style if style is linked, and linked of necessity, to unity and unification. Hence, it becomes the stylization of style. What has to be argued here is that the opposition between the particular as the idiosyncratic and the universal—style as universal—that allows for and sustains unity is not the opposition within which the question of style should be thought. The Hegelian setup in which all that seems possible is a relation between the universal and the idiosyncratic particular will be, in part, undone by elements at work in Benjamin's encounter with Jugendstil.

At this stage, however, there is a more general question: How is the reference to technology in passages of this nature to be understood? The question has to be posed first in relation to the distinction between *Werkform* and *Kunstform,* or more exactly in terms of what happens to this distinction once fashion comes to define both time and subjectivity. And then second, in terms of what happens to the distinction once the product—the object of technological production—becomes the commodity. These two points overlap, since while the movement to the commodity can and should be given its own sociological and historical account, the force of Benjamin's approach—and it is an approach that has to be, in part at least, wrested from the fragments of *The Arcades Project*—is that once viewed as a movement within time, and thus as the movement of a certain conception of historical time, the movement of commodification—understood as a passage of time—will sanction its own interruption. In other words, once the movement to the commodity is construed as temporal. then the language of the "dream" and of an "awakening" becomes that within which it becomes possible to think a "genuine liberation." Thinking, and here the activity of cunning comes to the fore, is linked to action as an actual possibility (or at the least as a possibility having actuality). This would, of course, amount to the signaled move away from political idealism. What it opens up is a politics of particularity.

In terms of the detail of Benjamin's engagement with Jugendstil's own complex relation to technology, as it is present in Konvolut S, the lead is provided by Benjamin's description of Jugendstil as "stylizing style par excellence" (S8, 2). This description locates Jugendstil in the tradition bounded by the relationship between technology and art. It then goes on to indicate in what way the technological is detached from functional concerns such that the interrelationship between *Werkform* and *Kunstform* that dominated earlier discussions of this topic vanishes due to the stylization of technology, in its essential differentiation from function. The engagement with technology therefore occurs to the extent that the possibilities of the technological, and with it the material, are refused. In sum, this is the ambivalence of Jugendstil. The ambivalence amounts to an indifference to materials.

What needs to be introduced here is the question of how the relation to the modern figures within Jugendstil. The description of Jugendstil that occurs in an earlier Konvolut (K2, 6) articulates the particular meaning that can be given to its conception of the modern. In a sense, therefore, it can be taken as providing, if not the answer, then at the very least the framework for an answer to the question posed in Konvolut S; namely, "How does modernism become Jugendstil?" (S11, 2). Of the many possible ways of understanding this question, here it will be interpreted as concerning the problem of how, within Jugendstil, does the modern figure. As such, the presence of the modern cannot be reduced to just the question of appearance—the answer to that question is linked to the stylizing of style (style becoming literal style and thus appearance)—rather, it is the question of time and interruption. And it should be added that this is the definite point; the one that will allow a connection to be drawn between style and time. Benjamin presents this latter relationship when he notes that Jugendstil "is the dream in which one is awake" (*er ist das Traümen, man sei erwacht;* K2, 6). Earlier in the same Konvolut, the interplay of dream and awakening is used to position advertising not as a sign of deception, but as that which could reveal the truth. Truth may be that to which one awakes. The passage is given below. Again, it is a passage that has already been noted. It returns now because it brings the concerns of architecture and fashion together. Both are linked—though the different nature of those links will retain their particularity—to the general question of the interruption.

> Fashion, like architecture, inheres in the gloom of the lived moment [*in Dunkel des gelebten Augenblicks*], belongs to the dream consciousness of the collective [*zähen zum Traumbewußtein des Kollektivs*]. It awakes, for example, in advertising. [*Es erwacht—z.b. in der Reklame.*] (K2a, 4)

"The dream consciousness of the collective"—reinforced by inhabiting "the gloom of the lived moment"—is awakened by advertising.[21] Here is a contrast between a form of awakening—the dreamt awakening (enjoining, of course, its own subsequent awakening)—and another in which the gloom dissolves. If there is an initial difficulty in Benjamin's argumentation, then it will occur in the contrast of these two formulations.

What is an awakening that takes place in relation to advertising? Two elements are at work in the answer. The first is a conception of collectivity that is estranged from itself. It is estranged by the reduction of the collective to the individual or to forms of individualization. More important, it is estranged from the realization of its own possibility. Advertising allows for a presentation of the collective, interpolated as such—as a collective—by the modern symbol. Such a collective, of course, cannot realize its own possibility. There is, as most commentators point out, an inherently utopian element in this formulation. Collectivity as a possibility is both announced and denied in the movement through the "gloom." The argument, however, depends upon the potentiality for collectivity being latent in order that the gesture of announcing and denial can be seen at work. It is not difficult to envisage elements of nostalgia in this formulation, even if it is nostalgia for a possibility whose actuality is yet to occur. The twofold movement of the awakening that is allowed by advertising has to be distinguished from the dreamt awakening of Jugendstil. How is this distinction to be understood? Is it literally a distinction between populism and its latent utopian potential and a type of false perception? Even if this literal interpretation were accepted, then it would still be necessary to argue that even within advertising, though more significantly in the possibility of a dreamt awakening, what was actually at work was time and thus the possibility of a reconfiguring of the time of the present—a reconfiguring that would itself be a reconfiguring of the present. With this shift in emphasis, potential would no longer have to be identified literally with that which allowed its presence to have been identified. Interrupting the literal would be cunning.

For Benjamin, Jugendstil is bound up with forms of sterility. When the straight line that characterized Schinkel's Bauakademie in Berlin becomes increasingly curved as the stylizing of the line in, for example, van de Velde's design for the theatre in Weimar (1904), it is not as though all that is happening concerns the line and thus its appearance. What is taking place is the complete stylization from the exterior to the interior in ways (forms) that are antithetical to the material with which—or perhaps in which—it is occurring. It is as though the presence of technological possibilities is being betrayed. The act of betrayal gives them an inwardness that defines their being present; i.e., their presence as appearance. The insistence on interiority (and Benjamin is, of course, right to recognize that with Jugendstil it is not as though there is any element that remains without style) means that in the end Jugendstil will lack a truly public presence precisely because its appearance is idiosyncratic. There is no opening into the realm of the public for the simple reason that the public—and the public realm as the domain of unity and unification—is not the universalization of the particular. This captures the strength of the position formulated by Benjamin in "Experience and Poverty" in which the concern is not the individual but the "poverty of human experience in general" (*SW* 2:732/*GS* 2.1:215).

From the realm in which the paradigm becomes the individual and the idiosyncratic, the countermeasure involves a recalling of the possible—though also impossible—community allowed by the commodity and by advertising. This is the community of the other symbol. And yet, of course, this is not an adequate counter-positioning. If Jugendstil is the dream of being awake, then one awakening comes from the link between collectivity and advertising. (This is the utopian impulse.) The other comes from the interpretation of that positioning of dreaming and being awake that involves a determined use of time. In other words, the suggestion is that the response to the simple positing of the new—the response that allows such a positing to be diagnosed and acted on—is divided between the utopian dimension of Benjamin's thought and the more complicated (because less worked-out) aspect in which modernity and its inception yield a temporalization of the political (and thus a politics beyond the vacuous demands of the utopian).

Indeed, it is possible to argue that this tension—or division—inheres in the question of style from the beginning. Once *Kunstform* is linked either to the symbol such that its unifying effect is immediate, or is attributed a semiautonomous status such that it achieves the same end within allegory, then the response to the object, and more significantly the conceptual determinations taken to the object, would always be bound up with the object's capacity to be unified and thus to unify. The relationship to technology that is implicated in both the development of fashion and equally, though clearly relatedly, in the process of commodification, is implicated in arguments about unity and collectivity. Turning this position around, what still endures as an open question is the extent to which, as a result, any thinking or rethinking of collectivity remains bound up within a politics of the symbol. While the symbols would shift, structures of unity are still determined symbolically. Countering such a result involves using the other element of technology and linking it more emphatically to the question of time. Such a possibility stems from the position that is inherent in the critique of Jugendstil. Part of the real force of Benjamin's argument is that Jugendstil betrayed the possibility opened up by the technology that it used. If the term "technology" is replaced within a reworking of this position—perhaps as a reworking of this position by the term "technique"—then a different possibility emerges. Elements of that other possibility will figure here in the guise of a conclusion to this chapter.

Final Openings

In his response to Oscar Schmitz's critique of the film *Battleship Potemkin* (1927), Benjamin argues the following concerning how the impact of the technological is to be understood:

> The vital, fundamental advances in art are a matter neither of new content nor of new forms—the technological revolution takes precedence

> over both. But it is no accident that in film this revolution has not been able to discover either a form or content appropriate to it. (*SW* 2:17/ *GS* 2.2:753)

The significance of this position is considerable. Rather than generalizing technology, what can be taken from this position is a claim about particularity and therefore about the specific. Once the move is to the particular—and Benjamin is attentive throughout this short text to write of particularity and of the "case-by-case"—then any monolithic sense of technology has to be replaced by the specificity of techniques. What will mark out specificity is as much the medium as it is the technique—or techniques—in question. Once emphasis is given to technique, then any insistence on appearance and on style as ends in themselves has to be repositioned. Furthermore, if any retention of *Kunstform* as unifying and as a unity is displaced in the name of an ineliminable cosmopolitanism that will define the modern, then conceptions of collectivity—be they utopian or not—will also be displaced. Collectivity—except as the purely pragmatic—only works within a politics of the symbol. (Such is the power and tribal nature of sport.) What particularity demands—a demand that is in accord with the move from the singularity of technology to the plurality of techniques and hence in accord with the need to respond to the interruption that the technological stages—is the recognition that the language of the appropriate must be stripped of finality and teleology. The appropriate must figure beyond the hold of any creative or political moralism. Hence, the question of style has to be posed in relation to techniques. As such, style will reappear beyond any simple oscillation between form and appearance. This is the style of modernity. Responding to its imperatives is to respond in its style.

If there is a final word that can be added to this analysis, then it has to involve a return to Hübsch's use of the question mark in the title of his pamphlet. If the answer to the question of style is not already known, then what are displaced immediately are any transcendental guarantee (the hope of idealism) and any effective retention of tradition (the hope of historicism). And yet, such a displacement does not generate simple particularity, as though all that remains are mere moments. In one of the most reflective comments in *The Arcades Project* on the status and form of the work itself, Benjamin identifies in one of the "First Sketches" what determines his undertaking.

> Of the rhythm of today [*Zum heutigen Rhythmus*] which determines [*bestimmt*] this work. Very characteristic is the opposition, in film, between the downright jerky image sequence, which satisfies the deep-seated

need of this generation to see the flow of development disavowed, and the continuous musical accompaniment. To root out every trace of "development" [*"Entwicklung"*] from the image of history and to represent becoming [*Werden*]—through the dialectical rupture between sensation and tradition—as a constellation in being [*durch dialektische Zerreißung in Sensation und Tradition als eine Konstellation im Sein*]: that is no less the tendency of this work. (H 16)

This is followed in the drafts with the passage that becomes N1, 9. Its opening lines are relevant here: "Delimitation of the tendency of this project in respect to Aragon; whereas Aragon persists with the realm of dream, here the concern is to find the constellation of awakening."

Identified in the first of these passages is a formal position that is determined by the needs of the present (what prompts the undertaking—in both form and orientation—are the "rhythms of today"). The terminology is of considerable importance. At work is the relationship between "rhythm" and music; and it should be added that their relation is importantly discordant. The contrast between film as containing the continuity of the discontinuous—a formal claim about film, and thus one often belied by its content—brings with it that which allows for a thinking of time and which is marked by interruption. (An allusion has already been made to the link between discontinuity and a "temporal montage.") What is staged, therefore, is the truth of time: what was identified earlier as a repositioning of the Absolute. As has been argued, what this means is that time is able to figure as a site of activity and therefore of political intervention. As the distinction between the jump cut as a sign of the discontinuous and the "continuous musical accompaniment" makes clear, what is at stake here is a difference that can only be adequately explicated in terms of a temporal disjunction. And yet, of course, it is not a simple disjunction, as is clear from the rest of the passage. This second half of the passage is more programmatic in nature. The program has to be read as connected to temporal concerns. The task generated by the truth of time is to undo the hold of continuity—a state of affairs whose possibility depends upon the truth of time. The distinction he draws is between "development" and "becoming." The richness of this as a philosophical distinction should not go unremarked. The distinction is between a conception of time as continuous evolution—perhaps one in which intervention and thus "awakening" could not figure—and "becoming" *(werden)* as allowing for interruption and thus not as a simple end in itself. Thus Benjamin's location of becoming, to use his term, as "a constellation in being." "Becoming," however, is neither eternal recurrence nor the repetition of the always the same. Becoming—that which allows for works and events to have an afterlife, and therefore a term, in Benjamin's formulation, that is bound up with the truth of time—is presented in terms of an interruption. The representation of "becoming" involves the "rupture" between what is identified as "sensation" and "tradition." These terms allow a further translation, this time in temporal terms. At work here is a twofold conception of moment. The rupture is a moment. The sensation occurs in a moment. Tradition is constrained to incorporate such "moments," since tradition is the continuity of these moments. What is being staged here is the possibility of a moment that tradition cannot incorporate. This is not pure spontaneity. It might be the "discovery" of content "appropriate" to a specific artistic form. What is fundamental is that the rupture is the caesura. It is the counter-rhythmic force. The twofold nature of the moment links the counter-rhythmic force as much to the tradition of Early Romanticism as it does to the need to see a link between "cunning" and the moment. The link defines activity. Moreover, it delimits the temporal nature of the political. Conflict is identified by the work of time. As such, of course, conflict has to be central to the historical process. Whether the reference be to Marx or to Lotze, the point endures. If humanity remains "mired in an uncivilized condition," then "how upon such assumptions can we be entitled to speak of one history of mankind?" (14a, 2). Now, while this is the case as a political claim, and moreover, while it is a claim that is fundamental to the development of a politics of time, it is also true that there needs to be a formal presentation of this state of affairs. The story of discontinuity demands its own formal presence.

Part Two : **Unforeseen Appearances**

4. The "Place" of Cosmopolitan Architecture

What then of the place of modernity? This formulation designates the site in which the conflicts marking the presence of modernity come to be played out. The question is direct and its stakes unequivocal. And yet the inherent difficulty concerns how such a predicament—the predicament both located and raised by the question of place—resides in the further question of how place is to be thought. What is an idea of place? If part of what has been discovered in working through Benjamin is the need to overcome the attribution of an idealized, if not essentialized, understanding of the language of thought, then the issue at hand concerns the possibility of thinking "place" beyond the hold of idealism, even of an ideality of matter. In order to respond to these concerns, the mediating figure in this context will be Georges Bataille, and specifically his engagement with materialism. What is wanted, after all, is a materialist conception of place.

As part of an investigation of what he termed "materialism," Bataille had two major concerns.[1] The first is that materialism not be reduced to an essentialism of matter. The second is that matter, and hence materialism, be held apart from any form of idealism. The specific form of this argument is central. Bataille writes:

> It is time now that when the word materialism is used it designate a direct interpretation, excluding all idealism, of brute phenomena, and not a system founded on the fragmentary elements of an ideological analysis elaborated under the sign of religious accounts. (*OC* 1:180)

What this means is that matter not be explicated in terms of an idea.[2] While this may seem to be no more than an abstract philosophical claim, it is, in fact, the challenge for a cosmopolitan architecture. In overcoming the possibility of creating matter that exemplifies an idea, what is created in its place is the material presence of an architecture that resists a conception of both symbol and ornament (elements of a possible though now putative unity of style) defined in terms of a singular idea. The resistance, perhaps refusal, cannot be simple disavowal. Rather, there must be a creative opening. The refusal of the symbolic, and the distancing of any comfort that symbols may bring, opens up the space of the cosmopolitan. Equally, they provide the space of the cosmopolis. This is the argument that needs to be developed. Terms such as "topos," "place," "region," "local," etc., only have real critical force once they are brought into relation to a cosmopolitan rather than a national mode of thinking. (This will involve more than a recasting of the nation in cosmopolitan terms.) The issues here are as much architectural as they are political or philosophical. Developing another mode of thing—a development occasioning its own interruption—necessitated another mode of thinking. Hence there is an inherent philosophical element. Bataille's undertaking to develop a nonidealist conception of matter has to be understood as a philosophical intervention, one with all the strengths and weaknesses that the term "philosophical" will carry in such a context.

Bataille and Architecture/Symbol

The entry "Architecture" in Bataille's *Critical Dictionary* (1929) opens up a number of important possibilities. As a text, however, its full force only emerges when, instead of working with the assumption that Bataille is writing on architecture *tout court,* it is recognized that central to the argument staged in the entry is the symbolic dimension of architecture.[3] Strategically, therefore, what this means is that a distinction needs to be drawn between architecture as an internal arrangement and delimiting of space, on the one hand, and the symbolic dimension of a building—perhaps even of a logo—within the urban field on the other. While it will be necessary to complicate this distinction, given that there will be important points of interconnection, it is nonetheless vital to hold on to it at this stage. As a point of departure, Bataille's concern is with the symbol; moreover, he will interpret moves against the symbol as symbolic. Both of these positions are advanced in the following passage.

> It is in the form of cathedrals and palaces that the church and the state both address and impose silence on the masses [*multitudes*]. It is clear that monuments inspire social wisdom and often even real fear. The taking of the Bastille is symbolic of this state of things. It is difficult to explain the movement of the crowd [*foule*] other than by the animosity of the people [*peuple*] against those monuments which are the veritable masters. (*OC* 1:170)

Perhaps the most striking element here concerns the "animosity of the people against monuments." The response to this animosity is not to create the anti-monument. Monuments become monuments—matter is realized as monument and is only able to be maintained as such—because they function symbolically. Monuments survive, as monuments, because they recall a founding event; more significantly, however, that survival is premised on their having a symbolic dimension. Rather than asking—what is a symbol?—the real question has to concern what it is that a symbol effects. In other words—what do symbols bring about? (Or more exactly, what are they intended to bring about?) When Bataille writes that "the great monuments [*les grands monuments*] are erected as dikes opposing the logic of majesty and authority against all the troubled elements [*à tous les éléments troubes*]" (*OC* 1:171), he underlines the power of the monument as symbol. At work in this "opposing" is the projected unifying effect that defines the symbol's modus operandi. The symbol unifies. And yet that projected unity may be far from benign. Unity becomes the imposition of order on the many. There are at least two ways in which the symbol projects unity. It needs to be added, of course, that precisely because the unity need not be benign, and moreover may have to be enforced, that unity is only present in name and thus as a type of construct. To naturalize unity is to forget its fabrication. At the present, the conflict must always be to the complexity of the intrusion of this nature. Part of this means that the resistance to the symbol has to involve a rethinking of unity. The destruction of symbols cannot be symbolic.[4] Symbolic destruction is always literal.

Order is imposed first by the symbol presenting itself as the locus of identification and, second, by ordering the urban field in a unified way. While these moments may overlap, they need to be treated in turn. In regard to the first, there is an important connection between symbol and ornamentation. In both instances, what is at work is a specific intention, namely, the construction of an object that necessitates identification. Leaving aside the question of success, once again, the basis of that identification is the expressive quality of the object. In expressing a singular idea and thus being regulated by that idea, the symbol grants identity while depending for its effective nature as a symbol on being a site of identification. The symbol demands a commonality that it sustains and reinforces. What this means is, of course, that the symbol presupposes what it demands. The link between symbol and ornament makes Adolf Loos's critique of ornament even more emphatic. His critique of "folk traditions" is not to be interpreted as an attack on popular culture. Rather, it is part of a sustained move to establish a link between modernity and the now-problematic status of style. However, it has to be linked in such a way that style has been wrested from a theory of expression. Note the following claim made in Loos's "Ornament and Crime."

> As ornament is no longer organically related to our culture, it is no longer the expression of our culture.[5]

The force of this claim is bound up with its explicit critique of "expression." What it opens up is a question—a question not resolved by Loos—concerning what is appropriate to culture. Does modernity have its symbols? If the modern is automatically identified with the contemporary understood as *this* historical period, where the conception of historical time involved is simply chronological, then it does have its symbols. And yet they are the symbols of global capital. They are neither national nor cosmopolitan. It will be essential to return to this point.

Once it can be argued that modernity is not the same as the contemporary, but that modernity involves a profound interruption within continuity—and this, as has been argued, is a claim as much about content as it is about historical time—then it is important to recognize that fundamental to the struggle to maintain modernity is the crisis of the symbol. It becomes a state of crisis precisely because the symbol has to refuse both a discontinuity on the level of content and an interruption within the temporality of chronology. However, it would be simple complacency to suggest that the crisis means that the modes of identification demanded by the symbol have gone. While on one level symbols may fade, the growth of nationalism is inevitably articulated via the symbol. Symbols—their presence and thus the possibility of the countermeasure that would entail, if not their absence, then at the very least the abeyance of any power that they might have had—define a space of activity. The unfinished nature of modernity is evidenced by this crisis. While symbols may have become strategic, what endures is the question of the relations between the symbol and a projected cosmopolitanism.

The second sense in which the symbolic dimension of architecture imposes order is in terms of the regulation of an urban field. Perhaps the most dramatic form of the symbol in this instance would be either the ghetto wall within the city or the city wall as delimiting a type of frontier. It is not sufficient to argue that these points are of mere historical interest because, for example, there are no longer ghettos; both possibilities have been rearticulated within a certain conception of the urban. The ghetto wall within the city and the wall of the city-state are not just moments of physical presence. The ghetto wall works because it confines and defines but does each on the condition of its permeability. It is, of course, a permeability that at any point can be denied. Permeability is the mark of the continuity of another sense of crisis insofar as it attests to the inherent instability of the wall's symbolic role. The ghetto wall has a twofold effect. In the first instance it divides two communities. The second is that such a definition then defines those communities in terms of an opposition; however, in so doing it homogenizes both of them. The centrality of permeability is that it allows for continuity on the level of the economic. As participants in the economic, the wall has to be crossed. All members of the city are the same—recognizing the power differentials with such a conception of the same—in terms of their participation in the economic. The importance of a daily act of reenclosure is that it announces the impossibility of maintaining such a conception of sameness outside any sphere other than that of the economic. Even then, participation in that realm is always a complex and highly regulated activity. The work of the ghetto wall beyond the physical act of enclosure is symbolic; it defines identity and holds it in place. The necessity that walls and borders be permeable indicates the possible fragility of the realm of symbols in relation to the economic. Reciprocally, of course, it indicates the extent to which economic sanctions necessitate the closure of borders. Such enclosures reinforce the homogenization of the enclosed. Equally, those enclosing are rendered the same.

Even though in most urban contexts, the forms of literal enclosure marked by the ghetto wall or the city gate may have vanished and therefore their organizing effect may have been dispersed, the site of that dispersal is the urban fabric itself. What is registered now is the orchestration of sameness that occurs through the proliferation of symbols. While symbols of national identity may have been displaced—or only able to be used in terms of an ideological recidivism that is condemnable in advance if thought to express any form of uncontested, or incontestable, essential unity—their removal is marked by a replacement in the form of international symbols.

It is as though, within an urban context, vernacular has been replaced by an international style. Vernacular and the international form part of an opposition that in their different ways hinder the development of the cosmopolitan. Interrupting that opposition and allowing for another spacing will have to take place in the name of the cosmopolitan. With the fabric regulated by the reiteration of symbols that have a unifying effect, there are still divisions and borders. It is as though frontiers and ghetto walls crisscross the city. Rather than attempt to erase them in the name of unity—a practice that would entail a specific urban practice—they could be taken as the site of an urban complexity. This is an important point. The logic of the ghetto wall entails either that it be maintained as such, or destroyed. Rather than assuming that this is the opposition that has to determine thinking and action, what has to be opened up is the possibility of articulating divisions and borders as descriptive of the urban fabric itself. Instead of assuming that these borders and frontiers mark points that generate and sustain identity, they would mark simple divisions and contexts where issues such as movement had to be renegotiated. Allowing for divisions involves an inherent realism. What is important is that complexity not be effaced in the name of an enforced unity. What has to be held in place is another conception of spacing. However, the question is how the spacing of the cosmopolitan is to be understood. Answering this question involves another opening, since what is being asked by such a question is another thinking of place and thus another thinking of "our" place. And who then are "we"? What has to be questioned, therefore, is the subject of the cosmopolitan. Another way of asking this question would be to address the concerns of the subject of modernity.

Nietzsche and Original Homelessness

Let us turn to Nietzsche—to that moment where the celebration of "racial impurity," impurity as an affirmative and positive quality, comes to be mixed with an evocation of the modern, in order to begin to identify the subject position of the cosmopolitan. This return yields a point of departure. In what follows, section 377 of *The Gay Science* sets the scene. Indeed, what follows could be understood as working through what is set in play by the opening words of this section. Prior to citing them and thus allowing that citation to work as the beginning, it should be noted that these words both open this section and then are repeated within it. Their repetition forms the opening of the final paragraph of section 377. The repetition of these words needs to be identified, not just because of the importance of allowing for the materiality of writing to play a fundamental role in the formulation of Nietzsche's position, but because that repetition announces the very problems with which any interpretation of this setting will have to negotiate. That setting is the complex relations—lines—whose intersections mark out the position of the subject of modernity.[6]

As a way of beginning, therefore, section 377 displays this subject. It begins with a two-word sentence; perhaps a two-word heading. Either way, the problems set in play by these words announce an immediate difficulty. Section 377 begins thus: "Wir heimatlosen" (We who are homeless / We homeless ones). The presence of this phrase in book 5 of *The Gay Science* allows for the initial suggestion that such a position is part of the general description of separateness that names book 5 itself: "Wir Furchtlosen" ("We Fearless Ones"). Fearlessness is not a position that pertains to the individual in its relation to its surroundings, as if those surroundings had an arbitrary location and an indefinite force. On the contrary, it designates a stand in relation to the given. Here the given is the understanding of tradition as that which is given in advance to be repeated. It determines action and thus yields conceptions of unity and disunity. It determines the tradition of place. The possibility of fearlessness, once enacted, would be the affirmation of homelessness. Here this has to be understood as involving the refusal of the determinations and identifications demanded by the symbol. This position has to be understood in terms of both a positive and a negative element. In other words, it is not as though the refusal of the symbol demands its destruction, or moreover that the elimination of the ghetto wall entails that "we" are now all the same. The situation is inherently more complex.

Homelessness—in the precise sense in which Nietzsche is using the term, and it is the demand for homelessness given that home would be identified with a certain tyranny of place, rather than the demand for shelter—necessitates courage. However, in this instance, courage is not the mark of nobility. Courage is a stand that is taken in relation to the question of the nation and of place, given the articulation of those questions within the structure of myth. (And this will be the case—namely, that this articulation will have occurred—even if the mythic status is either occluded intentionally or simply remains unnoticed.) The evocation of "homelessness" appeals not to the ancient, in order that it is overcome by the retrieval of that which would secure the home. Homelessness becomes a particular stand within the present that concerns an understanding of the present, a stand which by its very nature involves cultural practices. In this instance, what such an understanding marks out has to be construed as the complex interplay of affirmation and experience and thus a rethinking of a place that is determined neither by the reiteration of the vernacular—where vernacular is allowed both a literal and figural presence—nor the international. Allowing design such a role is not to give to architecture the task of enacting social change. Such a demand, despite being prescriptive, fails to understand what a politics of architecture would be like. What have to be thought through are the architectural and urbanistic correlates of an experience that dislocates. The virtue of the term "dislocation" is that it brings place—location—into a productive conjunction with a thinking of interruption; in other words, it stages the productive nature of the caesura as a strategic move. (This is the point at which a philosophical reflection on experience and the role of experience within culture coincide.) Dislocation is linked neither to violence nor nihilism. Rather, and in this context, it is the affirmation of an opening. What has been identified, therefore, is an opening—an allowing—other than those given by the determining constraint of the effective retention of either the national or the international. What this means is that dislocation becomes the precondition for another thinking of place.

Despite their difficulty, and precisely because of the link between dislocation—perhaps as a further translation of "homelessness"—and how place is thought, it is vital to trace some of the elements at work in Nietzsche's formulation of this now "modern" homelessness.

> We, who are homeless, are too manifold and mixed racially and in our descent being "modern men," and consequently do not feel tempted to participate in the mendacious racial self-admiration and racial indecency that parades in Germany today as a sign of a German way of thinking.[7]

While the nation is named, Nietzsche is unequivocal in his contrast between the German and the European—"we are . . . good Europeans" (*Wir sind . . . gute Europäer*)—such that simple particularity is no longer fundamental.[8] In its place there is a complex contrast; or perhaps, a contrast that brings with it a complexity that undoes any straightforward positioning in relation to the national question. Indeed, it will always have to be a question of how the nation is thought. Thinking the nation cannot be separated from the different and incompatible ways in which identity can be understood. (There is a broader argument here that ties a form of philosophical thinking to a concern with the cosmopolitan; namely, that philosophical thinking of the nation is already a thinking of identity and thus has to be approached on that level.)

The first element to note in this passage is the reiteration of a conception of national identity that is articulated within the structure of myth. It will be a myth that has been normalized and is thus lived within that conception of national purity that the actualization and thus the naturalization of myth demands. What is identified in the passage as "racial self-admiration" and "racial indecency" mark a specific thinking of the present. It is there in "Germany today." (Nietzsche's formulation is precise—"today in Germany" [*heute in Deutschland*]—and is linked to "a German way of thinking.") The relation between them constructs a specific home. It is, of course, the home to which the response is not conditioned by a simple form of identification or nonidentification. The evocation of "homelessness" marks a different form of nonidentification; one located beyond the oscillations set in play by the opposition home/homeless. What introduces difference here is that the contrast is not between different senses of belonging to a nation, positive or negative, but that the setting determining that response is given by the actualization of myth. Homelessness has its setting within that site to the extent that it starts from the position of recognizing that what determines the present, what generates points of conflict within it, is, to extrapolate from Nietzsche, the naturalization of myth. With the affirmation of "homelessness," therefore, myth is experienced as myth and not as natural. The demythologizing move, however, is not the work of critique, as though the subject position remains the same. Rather it is the distance that stems from the recognition of an ineliminable hybridity that works to define "modern man." This is the structure of recognition enacted by the reiteration, in the text, of the opening words: "We homeless ones" *(Wir heimatlosen).* Moving from designating a particular setup to its being that which allows the setup's particularity to be explicated, is the work of the opening's recitation. Being "homeless" will be that which affords the recognition that identifies—thus allowing for—its own proper position.

What is of central importance in this evocation of the "modern" is the link between it and hybridity and then how that link contrasts with the reiteration of identity—here national identity—thought within the structure of myth. The contrast of racial mixing and racial self-admiration is mediated by the latter being described as "mendacious." Recognizing it as such involves seeing the history of the race—either as the pure race, or as that which is given within the posited singularity of national identity—as the history of a lie. Not the lie as opposed to truth, as though it were merely a question of choice, but a lie that comes to be recognized, and therefore which comes to be experienced as such. It is thus that the counter to the lie involves neither the simple positing of truth nor the outside. Homelessness has a twofold determination. It involves in the first place the recognition of a displacement occurring such that the interruption of myth stems from the recognition of myth as myth. (Interrupting thereby the linear narrative of temporal continuity in which myth would have continued to be actualized—becoming actual within its naturalization.) The second is that this recognition is sustained by, if not itself given by, an impurity within the "us," the unity that "we" are—remember that Nietzsche writes, "We homeless ones," "Wir Heimatlosen." The mixture that marks the modern cannot be understood as an addition, as though it were a mere predicate that comes to be given to "modern man." Such a state of affairs maintains the subject as that to which different predicates can be applied but which, nonetheless, unfolds within a sequential conception of historical time. Here the claim is different. The important point is that hybridity, impurity, homelessness are original states marking the subject of modernity; perhaps more accurately, they define the possibility of thinking the subject of modernity as itself containing—containing by having been constituted by—the preconditions for demythologization.[9] Here is the point at which a thinking of the cosmopolitan can begin to emerge. It will be necessary to return to this formulation in order to make more precise the conception of the origin at work within it. In other words, what demands further explanation is the sense of origin that is deployed in the description of hybridity, impurity, and homelessness as original. Unfolding this origin will provide a way of linking more directly ontological concerns—those pertaining to the being of being human—and ostensibly historical ones.

What then of the opening words: "Wir Heimatlosen"? As a beginning, the full range of the word "Heimat" (home) should be allowed in this description. The counter to the home is not homelessness, as though that were a simple either/or. Homelessness marks the modern. However, it does not mark modernity as such. There is the countermove; the one evoking what is taken to be home. Nietzsche identifies this setup with great acuity. It is that which "parades in Germany today as German thinking." What parades has been recognized as a mere parade. This is the moment of demythologization—the interruption that occasions. However, the countermove involves a different set of practices, and thus different subject positions. In other words, another politics of time. Opting for the mythic—there are other forms of identification also operating within the hold of continuity—not as though it were myth but with its having been naturalized, myth having become nature, is already to have identified with traditions that seek to sustain a singular conception of national identity. Maintaining this position, holding to it beyond the sway of irony, is modernity's other possibility. In other words, there is the twofold movement of dislocation—announced here by Nietzsche's linkage of homelessness, racial mixing, and the modern—and the continual attempt to efface that complex setup in the name of unity. Any attempt to enact that possibility would necessitate the symbol.

What this entails is that "homelessness" is at home within modernity as that other move that will always reveal the truth about modernity. Its truth is that it is the disruption—the dislocation—bringing with it the countermoves of myth, historicism, and the positing of unity and the specific sense of singularity that seeks to efface dislocation. As has been argued throughout the preceding, the particularity of modernity is that one brings the other with it. The struggle to maintain particularity enjoins strategies displaying dislocation, the affirmation of the interruption that delimits modernity's ineliminability. The opening declaration "Wir Heimatlosen," its subsequent recitation as establishing the productive link with the modern, has to be interpreted in the first instance as this display. The repetition, therefore, in opening up the question of the home, in staging the place of that question, is at home within what is being staged. However, because being at home is not a singular position—hence the home of homelessness—being at home allows for a critical recognition; moreover, is that critical recognition? In refusing the nihilism of simple oppositions, the affirmation of "homelessness" and the subject positions that are staged by it work as an affirmative possibility that in coming to be recognized—and here the expression "coming to be" is intended to mark the process of recognition and thus experience—opens up the twofold determination of modernity. Once opened up, opened up by this particular affirmation, then both the affirmation and what is displayed by it become the work of truth. A determining setup is displayed through this work. Due to the impossibility of an absolute escape from determination, in this instance truth's work is inextricably bound up with homelessness, impurity, and hybridity as original states.

While there are other elements that would arise from a more detailed consideration of section 377 of *The Gay Science*—perhaps the most important would be the nature of the link between "homelessness" and "fearlessness"—nonetheless, what has emerged thus far generates a specific task. The task is twofold. In the first instance, there is the question: How is the possibility of original hybridity, impurity, and thus homelessness to be thought? And second, there is the further question concerning how such thinking is to be staged. This second question does not refer to the instrumentalization of the philosophical or the theoretical. Rather, what it addresses is the possibility of an architectural activity that would be sustained by the affirmation of original homelessness. Original homelessness understood as a thinking of modernity will allow for the development of what can be called an architecture of the *other place.*

The Other Place

One of the problems inherent in any critique of tradition—in which tradition is understood as the naturalization of chronology—is that the identification, let alone the projection, of its other becomes an insistent problem. As has already been suggested, the distancing of the vernacular and thus of the reiteration of the unifying force of the symbol does not mean the incorporation of an international style in their place. What stands in the way of the international understood both as a style and as the internationalization of the symbol is the local or the regional. The local is neither the national nor the specific place of the international within a confined domain. What these terms designate—and there is no real reason to identify a real difference between them—is the "other place." Alterity is not defined as a variation; alterity marks difference. Thus, the question that arises is how the alterity in question is to be thought. What is an other place?

The other is here. However, it is not the "place" that is simply given. Nor is it the place that is imposed. What allows for it to be other is that it is both same and other simultaneously. This simultaneity in which a relation of distance is established becomes the way of understanding the local when the local is no longer simply part of what is already given. Understanding a conception of the local as determined neither by the national nor by the international necessitates beginning with an identification of the local with the cosmopolitan. This will have to be the case even if that identification is yet to be substantiated. Part of the argument that would need to be deployed in order to establish such a position would flow from the moment of overlap between the claims made by Nietzsche concerning an affirmative conception of homelessness, on the one hand, and the severance of style and appearance and thus the crisis of the symbol on the other. The point of overlap will involve a cartography of the cosmopolitan and thus a regeneration of the urban field. What flows from the crisis of the symbol is that sites of identification—monuments—no longer organize the urban fabric except in either pragmatic or strategic ways. Monuments become part of a national heritage and thus lose their capacity to organize outside of those instances—and there may be many of them—in which the demand for continuity is precisely that which organizes. The reason why this can be described as strategic or pragmatic is straightforward. Such organizational imperatives are demands. Demands made in the name of identities and conceptions of continuity that, if they are not deeply contested, then are highly contestable. This existence of the ineliminability of contestation opens up the place of homelessness. It opens it up because it defines it.

Allowing for the impossibility of symbols to organize the urban field is not a form of nihilism. The nihilistic gesture would be to allow the impossibility of the vernacular to open up the space of an international style. (It needs to be added immediately that the style of this "international style" is more complex than the simple inscription of dominant forms in random contexts. The history of any claim about international style will have already borne that out.) The key to understanding the local and therefore to understanding the identification of the local and the cosmopolitan lie in the presence of contestation. What contestation means is that both form and function, in addition to the effect of building within the urban field (if not the field itself) in terms of its diagrammatic organization and its appearance, are negotiable. On one level this will have always been true. However, what would be taken to have ruled such negotiations would have been the presence of existing forms that, once construed symbolically, would have had a determining effect on all of these elements. Moreover, the singular determination of the symbol would have been reiterated within and therefore reinforced by the singular determination given to how function was understood, even if the form given to that function seemed to be readily adaptable. Again, the point needs to be made that it is not as though this contestation is always resolvable in what could be described as regional terms. For the most part it is not; hence the presence of the contestation. In the same way that questions of identity are still caught within the confines of a conflict between essential determinations on the one hand and cosmopolitan ones on the other hand, the same is true in regard to the architectural. Essential determinations refer as much to the national as they do to the international. The local or the regional will always serve to interrupt that opposition. Once contestation is allowed to be central and, therefore, once the particular is allowed precedence over the general or the universal, then there cannot be one style that is appropriate to the cosmopolitan. Hence, the truth of the cosmopolitan is that it emerges from the two moments that characterize the advent of modernity; namely, the severance of style and time and the effective—hence productive—presence of an inaugurating interruption. (Interruption then becomes the motif in which to think both relationality—i.e., in terms of difference—and through connectedness—i.e., in terms of separation and connection.) In sum, the absence of a determined style provides the conditions in terms of which judgment is possible. The site of judgment becomes the site of contestation. Once the conditions allowing for contestation are eliminated, and therefore once negotiation between form and function has been refused in advance—either through the retention of the vernacular literally or in terms of the symbol, or via the putative ubiquity of the international—then the possibility of judgment is no longer at hand. Judgment occurs once there is a decision within a range of possibilities that were open from the start. Judgment is bound up with the cosmopolitan.

Rather than prescribe a way ahead, a return to place has to occur as part of a rethinking of what is entailed by place. Place is neither empirical location nor the expression of a generalized and all-encompassing idea. The space of the cosmopolis is the continuity of a complex spacing without end. The "without end" marks the need for the continuity of experimentation. However, it is a need that has to situate experimentations within the continuity of the struggle to maintain the dislocation yielding modernity. What continues to need to be maintained, therefore, is the "homelessness" identified by Nietzsche. Holding to it in the name of modernity means maintaining the continuity of the struggle to create and re-create the cosmopolitan.[10]

Equally, what this entails is another materialism. As such, what returns is Bataille's reworking of that term. The force of his position is that in eschewing both the reduction of the material to the empirical on the one hand, and the regulation of the material by a generalized and singular idea on the other, what is opened up thereby is the possibility of a conception of materiality that is no longer expressive. Materiality would be defined by a type of internality, even though matter was still constrained by its being architecture. This would not be to claim that matter does not signify in terms of its exteriority. The claim is that what is signified cannot be reduced to the expression of a single idea. As such, matter would have distanced its symbolic dimension. (This will be part of the argument that will be pursued in chapter 5 by working through specific architectural projects.) And yet there cannot be a complete reversal of this position, since the risk with pure interiority is that it may become formally abstract. By "formally abstract," what is meant pertains to built form and not to an abstract diagram. Abstraction in this specific sense refers to the simple ubiquity of form. A programless form as a projected then realized end. The presence of this risk means that the opposition between the vernacular and the international needs to be understood as more complex. Part of the complexity, perhaps that which defines it as a complex, is the articulation of different types of repetition. Not only is there a repetition on the level of the symbol; there is also a repetition on the level of form. In both instances, matter is regulated by and thus becomes the expression of an idea. Matter construed in this way is idealist. Countering this idealism cannot take place by abstraction. Abstraction becomes an instance of architectural nihilism. Interrupting the repetition of the symbol and the oscillation between the ubiquity of international style and mere abstraction is the place of the cosmopolitan. The architectural, as well as the political, task allowed by this possibility is to maintain the place of "allowing." What make this possible are the interarticulation of the temporality of the "without end" and a conception of matter that no longer works with the constraint of the idea.

Repositioning Regions

If there has been a sustained attempt from within the discipline of architecture to develop a thoroughgoing and critical conception of place, then it is to be found in Kenneth Frampton's arguments concerning "critical regionalism."[11] In the guise of a conclusion, a connection will be made between Frampton's argument and the conception of cosmopolitanism adumbrated thus far. Part of the argument will be that what remains unexamined in Frampton's argument is the acceptance of a Heideggerian version of human being that informs his understanding of "earth."[12] However, the point of departure is not Heidegger per se, but the way these elements derived from Heidegger figure within Frampton's critical regionalism. The critical nature of this theory of regionalism lies in what he describes as an "arrière-garde position."

The overall argument is presented in the following terms. A critical arrière-garde has to remove itself both from the optimization of advanced

> technology and the ever-present tendency to regress into nostalgic historicism or the glibly decorative. It is my contention that only an arrière-garde has the capacity to cultivate a resistant identity-giving culture while at the same time having discreet recourse to universal technique.[13]

The force of this formulation is that it signals the refusal of both the vernacular and the international. In addition, the option of postmodern architecture—identified by Frampton as the "glibly decorative"—is also ruled out. This position allows a connection to be drawn, as he implies, between Jugendstil and postmodern architecture as differing permutations of the history of ornament (a point pursued in greater detail in chapter 5). What is opened is a return to place that takes note of what is discarded in the process; hence, the critical dimension. For Frampton, this return is positioned on the one hand by an acceptance of what he terms the "ubiquitous placelessness of our modern environment," and on the other by a positing of what after Heidegger is described as a "bounded domain" in order that the latter rectify the former.[14] The way this occurs is by becoming sensitive to the specific demands—ones as much climatic as cultural—that given places require. Techniques should be adapted to allow the local to be evinced. However, this is not an argument for a token display of local conditions. Display will inhere in the building's construction. The relationship between experience and construction means that locality and function are within built concerns from the start. (For Frampton this then becomes—in borrowing from Boetticher—an argument for the centrality of tectonics.) This decentering of the visual in the name of the tactile knits critical regionalism to a form of architectural experience rather than positing architecture as a simple visual performative. He can conclude the argument that

> the tactile and the tectonic jointly have the capacity to transcend the mere appearance of the technical in much the same way as the place-form has the potential to withstand the relentless onslaught of global modernization.[15]

Place emerges, therefore, as the site for a type of attunement. The question, of course, is how this attunement is understood. With what should there be an attuning? It is instructive here to look at one of Heidegger's formulations of the predicament of homelessness. In "Building Dwelling Thinking," Heidegger argues that

> however hard and bitter, however hampering and threatening the lack of houses remains, the proper plight of dwelling [*eigentliche Not des*

Wohens] does not lie merely in the lack of houses. The proper plight of dwelling is indeed older [*alter*] than the world wars with their destruction, older also than the increase of the earth's population and the condition of the industrial workers.[16]

What is being staged by the argumentation of this passage—and it is emphatically present in the use of the term "older" to describe the predicament of homelessness—is an attempt to define the loss of place beyond any argument concerning modernity. Moreover, homelessness is descriptive of the condition of human being. Its relation to place, therefore, will always define the region in terms of the particularity of human being and thus this "older" plight. In a sense, therefore, place remains untouched. It locates that which defines human being and allows for the transformative effect of that recognition. Place may be transformed in the process; what endures, however, is the conception of human being defining it. Here the comparison with Nietzsche is vital. Nietzsche's evocation of homeless is affirmative. (There is no sense of loss, but of an overcoming; nostalgia would cede its place to joy—hence *Die fröhliche Wissenschaft.*) The refusal is not just of the national and the international; equally, what is refused is the possible presence of an older plight defining—defining essentially—human being. Place therefore becomes the site in which this affirmation is practiced. Thus place is transformed in a fundamentally different sense.

Critical regionalism's relation to place accepts too quickly the description of "placelessness" by identifying it implicitly with Heidegger's description of human being on the one hand, and on the other with the ubiquity of international style (both as a series of techniques as well as appearances). What is left out is the possibility that there is a productive sense of homelessness. Were this to be accepted, then a different opening emerges, one sustaining another direction for criticality. One way it might be pursued would mean that the critical could figure as much in how program was understood as it could in architecture's relation to the forces of the national and the international. This is the sense of the critical that is necessary for a cosmopolitan architecture. If modernity demands forms of affirmation, then so will its architecture. The question that endures is, of course, in what style should it be built?

5. In What Style Should We Build? The Style of Cosmopolitan Architecture

The task here is to respond to the question of style by engaging with a number of different architectural projects. They form a sequence only because they stage the problem of architecture's self-image—given the separation of style and appearance—with great acuity. The architectural figures cannot be reduced to mere tectonic possibilities. What matters is the situation of architecture; the complexity of its site. In more general terms, the contention is that to the extent that style is brought into consideration and once bound up inextricably with the question of modernity, what then figures within any concern with style are time and place. Moreover, given that the term "cosmopolitan" can be viewed as a site of resistance to the effective presence of the opposition between the national (understood in architectural terms as the vernacular) and the international, then what has to emerge is another set of possibilities. Moving from one architectural form to another is not an argument for stylistic eclecticism. (Nor, on the other hand, is using the term "style" as a covering term evidence of stylistic generality.) On the contrary, both can be taken as evincing the founding split between style and appearance that marks the modern. That split, occasioned by modernity, makes the question of judgment demanding in both senses of the term.

As a beginning—and consistent with what has been argued thus far—the time in which the question of architectural style is to be located necessitates recasting historical time in terms of the continuity of the discontinuous. However, precisely because of overcoming a conception of time as smooth and homogeneous, and in which all occurrences are intricate versions of the same, it can also be argued that complex temporal modalities come to inhere in particulars. Hence, the centrality of time opens out beyond a simple identification with historical time (even if the latter privileges interruption of continuity). For example, the distinction between the immediate and the mediate—a distinction that will play a central role in accounting for the operation of different senses of surface—is a temporal one. Equally, time marks the nature of the distinction between the image as a still and the dispersal of the stilled image within the film. Again, deploying a conception of temporal difference will allow a productive distinction to be drawn between the plan and the implicit urbanism of certain buildings. In more ways than one, therefore, a concern with style establishes the problem of the relationship between architecture and time. The problem is multifaceted. In part, it lies in the complexity of time. In fact, the concern should be with the times of architecture. The plurality of times will have an important effect on the architectures involved.

The second element that plays a fundamental role in how style is to be considered is place. Sites and place occur within narratives. (Time reappears, therefore, as central to the construction and reconstruction of those narratives.) The narratives may be histories, they may be stories, they could be maps or even site plans. In every instance there will be the feint of neutrality. However, it is not as though the neutral needs to be counterposed to the impossibility of neutrality and therefore to the already predetermined. The problem of place—and thus of site—has to do with how lines are drawn and borders established. The argument is that lines of exclusion and inclusion traverse sites; once it can be suggested that a sense of anteriority pervades the present and that both can be located in a site—perhaps both have to be located as such—then the site becomes the intersection of lines; ones with different chronological dates, ones bearing different political effects and which mark the possibility of different histories. To the extent, therefore, that it can be argued that any site is already configured by the intersection of these lines, lines that are as much virtual as they are actual—and that acknowledging their presence becomes a way of determining how site and thus place are to be understood—then what all these considerations involve and what in the end they will demand, is that place no longer be thought in terms of the opposition national/international. (Such terms have an essentializing quality on what they are taken to designate as well as on the "place" on which that designation is then acted out.) Such an exclusive and limiting determination is precluded by definition. Equally, it will mean that the construction of a site—and therefore a sustained thinking of place—cannot be given mere literal presence. Once place becomes the locus of an intersection of lines—the importance to be attributed to one set of lines rather than to another will always be a regional concern (not pragmatic but regional)—the question of style will inevitably bring a configuration of place into play. Again, it is essential to note that it is one possible configuration, precisely because the differential quality of lines means that their interruption—an interruption that allows form to occur—in breaking with the literal, breaking by marking the impossibility of a reduction to the literal, will always have a form. In the same way that there cannot be a single or simple national history—the history told from a mere description of what would have to have been a single line—the intersection of differential lines will have a generative quality that cannot be restricted to one form. It is in terms of the interplay of these lines that it becomes possible to locate the impulse behind any research leading to a cosmopolitan architecture. The cosmopolitan is not just creating something different in the same place. Rather, the nature of the place changes and is changed as part of another conception of production. Action activates possibilities or potentials.

As has already been noted, arguing for a cosmopolitan sense of place starts with the recognition that any site is already overdetermined. The crossing of different lines is a spatial description. Equally, however, it is temporal. Different histories intersect by occurring within different time frames or by generating different temporal possibilities. Their copresence accounts for the complex temporality of the present and allows the cosmopolitan to be an always already possibility within the present. (The cosmopolitan is not a state of affairs "to come" in the sense that it is not a point to which progress is made; it is the present's potential.)[1] The lines of movement, flows of trade and capital, for example, which from a certain perspective account for the origin of towns will continue to work through the site.[2] However, the dense knot created by these lines, a knot that becomes the town with its own infrastructure, will operate on different temporal and spatial orders. The displacing of sites or their repopulation occurring because of the interplay of movement and settlement—the latter always being a precondition for the former—creates a sedimentation in which both temporal and spatial concerns continue to interact.

The Question of Style

From the earlier consideration of style, particularly as it was first formulated by Hübsch in *In What Style Ought We to Build?* it should now be clear that the attempt to break the determining hold of imitation is a recurrent motif within the repositioning of style as a question. Equally, it is an inherent part of the formulation of that question in terms of the separation of style and appearance. As has been argued, this separation occurs in its most emphatic form in those conceptions of the modern that ground its inception in an inaugurating interruption. (This rethinking of the present has to be understood as a conception of modernity.) Hübsch's pamphlet finishes with the following formulation:

> In every case buildings logically designed in their basic elements will rank much higher as works of art, even with the most infelicitous decorations, than the most exact imitations of ancient art [*als die getreuften Nachahmungen der Antikte*].[3]

Due to the centrality of the opening made by Hübsch's pamphlet to this more general reformulation of the relationship between style and time, it is worth reiterating that at the center of his argument is a twofold move in which the activity of the imagination comes to be championed at the same time as a simple miming of ancient styles is put to one side. As such, style came to the fore as a question. However, coupled to its presence as a question is the necessity to control the excesses of the imagination. In other words, what this early formulation of the question of style stages is the need to avoid the idiosyncratic. The latter is present as a possibility precisely because the idiosyncratic is one of the eventualities that cannot be precluded once there is a break between the present and a sense of continuity that would have been guaranteed by imitation. Boetticher's contribution to the debate, in a pamphlet written in 1846, reiterates the argument against defining "style" in terms of individual aspirations. However, neither that move nor the redefinition of style in terms of tectonic possibilities can be taken as ends in themselves. They are tied up with another element. The question of style is the national question. As was noted, for Boetticher, in regard to style, "only a whole nation can cause its inception." Moreover, a style may take an "epoch for its development." Not only is style tied to the question of the nation, it is equally part of the nation's natural history. Even recognizing the artifice of this nature, this formulation ties together style, the nation-state, and the temporality of historicism as one configuration with the modern. That it is not the only one and that it is in conflict with a temporality thought in terms of interruption are the defining elements, as has been argued, for an account of modernity.

What emerges, therefore, is a complex state of affairs in which the break with imitation allows for the national question to be part of modernity.

In other words, the impossibility of maintaining continuity through mime, and of there having to be a break within natural time—a break that constitutes the modern—once articulated within a concern with style, become insistent forms of the national question. While Boetticher is right, for a given moment, to connect style and nationhood, the break with imitation allows for style to have a projective if not futural quality. It may be, therefore, that a certain use of style and ornamentation will project—in the sense of aiming to construct—a nation's self-conception. (Not the nation at hand but the nation to come.) The monumental architecture that projects a future rather than celebrates a past should not be seen as an aberrant possibility, but as another way through the predicament of modernity, once the latter is understood as a break. (The argument is straightforward. Were it not for the break, then attempts to establish or reestablish national identity through style would have been unnecessary.) Once a return cannot be made, and once continuity is no longer an option, then the question of how to continue cannot be avoided. This is the inevitable and now unavoidable locus within which the question of style continues to emerge. Moreover, it provides the setting in which the work of the Slovenian architect Jože Plečnik (1872–1957) can be located. The importance of his work is twofold. In the first instance, it has this projective quality in relation to the national question. Rather than seeking to confirm and therefore to consolidate an identity, its aim is to establish one. Second, it marks the final possibility for a sustained architecture of the idea; i.e., an architectural formation where the elements are given coherence and identity by their organization, both structurally and visually, by a regulating idea—for example, national identity—that is necessarily external. The idea will give style an appearance, but it will be an appearance no longer governed by the logic of imitation.

Plečnik

Plečnik returned to Slovenia in 1922, at the age of fifty, after having worked initially in Vienna, and in close conjunction with Otto Wagner, and then in Prague on redesigning Hradcany Castle.[4] His return was bound up with a form of nationalism. However, it was not the form that sought to rediscover a mythic past. Rather, it was to redefine Slovenia in terms of a fabricated future. Architecture was to play an instrumental role in this fabrication. His project was to shift the gaze away from a concern with either the Austro-Hungarian Empire or Pan-Slavism—and thus a conception of nationhood defined by the geography of that region or a more extended though less precise sense of ethnic identity (one necessitating an eventual identification between ethnicity and land)—by moving it toward the Mediterranean (a construct of the Mediterranean). For Plečnik, what became important was an attempt to redefine Slovenian national identity by the use of symbols and ornamentation—a close interconnection of modernism and the resources of style. What is going on within the actual built works, as well as urban studies, is the sustained architectural attempt to establish what could be described as the work of a predominating idea as that which governs either architecture or the plan. This is the place where architecture is defined by an idea. Idealism can construct a conception of architecture and a related practice, as much as it can produce a theory of art and architecture. Plečnik wrote of "the inner nerve of art," and while there is in his projects a concession to materials and cladding and thus a response to the imperatives imposed on architecture by the work of both Boetticher and Semper, for the most part that "inner nerve" was not expressed in terms of a concern with materials—let alone with material functioning to realize certain effects—but in the manipulation and deployment of symbols.[5] The dominant use of symbols almost subordinates the role of materials and thus of tectonics in establishing the relationship between the idea and the architectural object. (This position can be readily identified in the interiors of his buildings. For example, in his 1938 plan of the stairway for the National University Library, it is the symbolic presence of columns and statues that creates such an effect.)[6]

While there are many examples indicating the way in which ornamentation and style are used in order to project a conception of Slovenia and hence Slovenian identity as defined by the Mediterranean, two will suffice. Both are located in Ljubljana. The first is the construction, by Plečnik, of bridges across the River Ljubljanica and which are known as Tromostorje (Three Bridges). They were built in 1930 and 1931. The second concerns what could easily be described as his major work in Ljubljana, the National University Library (which opened in 1941). In both examples, what needs to be noted is the way a possible stylistic eclecticism is held in place by the overarching idea of the Mediterranean.[7] It is this overarching idea—the idea as the unifying factor holding the constitutive elements of works together—that predominates.

The bridge project involved the addition of two bridges in order to replace the original one built in 1842. The bridges not only provided a processional passage from the new to the old city, they resolved a number of problems caused by the irregular spatial axes of the Square of the Virgin Mary. (The urban project, of which the bridges form a part, is also another domain in which Plečnik rethought the city of Ljubljana's relation to the Mediterranean.)[8] The original bridge had an iron balustrade. While that bridge was kept—even its dedication and date were retained—the iron balustrade was removed. The addition of a concrete balustrade to the old bridge established a visual continuity between it and the new ones. The continuity, which is there in the appearance and in the coloring, effaces the temporal disjunction between the old and the new. However, this on its own is a minor consideration, since what is more important is the way in which the visual simultaneity of color and appearance—perhaps what would have been called style, were style and appearance always the same—works to recall the Rialto in Venice. While this is consistent with additions along the River Ljubljanica in that there was a sustained attempt by Plečnik to open up the river, what is of greater significance in this instance is that the division between the two parts of the city is formally resolved in terms of a single unifying idea. An idea that is maintained and held on the level of appearance. The recall of Venice is there in the bridge's appearance. And yet this is not mere appearance, since presence and recall cannot be disassociated from use. Indeed, it could be conjectured that it is the utility of the bridge—utility as necessity—that allows the constancy of that recall to be maintained. This is a significant point, as it indicates the extent to which what appears to be decorative is not in excess of use.

The National University Library contains many elements that warrant detailed attention. Rather than take up the building as a whole, comment will be restricted here to the facade as viewed from Gosposka Street. Five elements cohere. What is of interest is what enables them to do so. In the first instance, the building itself appears as an Italian palazzo. Both the overall envelope and its presence in the urban context bear an important visual relation to the architecture of Italy. Next, the stonework on the facade, which recalls national building traditions, also interrupts the smooth facade. More significantly, it calls attention away from the flat facade that characterized the modernism prevalent in both Vienna and Prague. While modernism as a style is therefore both announced and distanced, it is equally the case that the presence of the vernacular can be accounted for in a similar way. In addition, though as part of the facade, there is a single Ionic column that dominates one corner. The final two elements are the Etruscan vases above the doorway and the statue by Lojze Dolinar of Moses between them. (Many commentators link the statue not just to Plečnik's religious commitments, but also to Moses' destruction of idols; objects should have a spiritual value.) The statue is, however, more complex than its religious dimension. Its evocation is not of Moses but of a Moses in the style of Michelangelo. Italy is retained but invested with a religious dimension, or a religious dimension is retained without the guise of northern European Catholicism (let alone Protestantism), but with the full aesthetic force of Italy and thus of a constructed Mediterranean civilization.

The move to the Mediterranean provides the line of coherence between all the other elements. The Etruscan vases evoke what for Plečnik was an important link between Slovenian and Etruscan culture. The single Ionic column is not intended to refer to a specific site. While the internal movement within the library from the dark to the light may have recalled a certain conception of knowledge as illumination—an architectural staging of Plato's allegory of the cave—the single column does not evoke, for example, the Temple of Athena Nike on the Acropolis. Style while appearance—not an appearance determined by the logic of imitation, but by an architectural idea—has ubiquity rather than specificity. There is a proliferation of columns within Plečnik's architecture. Each one is intended to open and sustain a link to either Greece or Italy. While the columns will always be specific, their link is invariably that which stages the nonspecific, because constructed, Mediterranean. The facade as a whole—work on which Plečnik was assisted by Edo Ravnika—contains complex elements that are on one level incoherent. What coherence they have is provided by the idea of the Mediterranean, and here that means the Mediterranean as the projected source of Slovenian national identity.

There is a further element that is also part of the process, one whose aim is to establish an identity through projection.[9] (A project that held the disparate together.) That element is, of course, place, the site of architecture. Projection, once articulated within the ambit of a unifying single idea, has a determining effect on place. That place becomes the nation where nation comes to acquire—in fact or merely as an aspiration—the projected identity. Once place is given such a determination, this will mean that the inclusions and exclusions that determine any national history are then integral to the construction of site. What are eliminated as part of the same movement are the traces of complexity that are, or perhaps were, part of the history of that place. What is being unified at the same time is the terrain. The site is given an identity that refuses any possible complexity. As such, what are then unified are the disparate elements that made up Ljubljana. There is an obvious corollary here. Once it can be argued that the single idea that drove a conception of object as the enactment of a single organizing idea is no longer appropriate, then not only will the history of inclusions and exclusions have to be reconsidered, but that reconsideration will generate another conception of site and thus of place. National history as natural history will have been "denatured," thereby generating not just another history but another place. What this means is that in renegotiating history, there will always be the literal place—the site that is measured and determined by the conventions of mapping—nonetheless, it will also be the case that henceforth such a place will come to be invested in and traversed by lines and determinations that cannot be reduced to literal presence. (Literally the same, though now with the release of its inherent potentiality.) Once site is allowed that complexity, it no longer forms part of the nation and thus a nation's natural history, if nation is understood as either having or acquiring a synthetic unity. Site therefore becomes a locality and its concerns become regional. The local and the regional become lines through the nation.

What allows for the move from this moment is not a critique that operates on the level of ornamentation—as though all that was of concern is ornament—more is involved. What emerges specifically is the impossibility of maintaining a conception of national identity, delivered through the operation of symbols and unified by the idea of the Mediterranean. This is not to argue that nationalism is impossible within the modern period. Indeed, this is not the case, since the relationship between national and style is a direct consequence of the modern period. Rather, the argument is that the unity given by the idea is no longer operable. While this may open up a proliferation of possibilities articulated on the level of "style"—in fact, it is possible to view the front facade of the National Library as postmodern precisely because it brings together the modern, the Ionic, and the Etruscan—there is a more significant occurrence. The removal of the unifying idea means that the version of modernity that linked style and the national question is no longer an option. The insistent question therefore is the following: if an idea cannot organize style, then how is style to be understood? It is important to move with considerable care in developing this argument. It has to be remembered that the force of Plečnik's position was not the linking of style and the national question in terms of a given possibility, but in terms of a projected possibility. Style had an instrumental quality. Style would have opened the future. In order to take this a stage further, a distinction needs to be drawn between a proliferation of symbols that takes place without an organizing idea, without project, and therefore where there are symbols simply as decoration or ornamentation, in other words, postmodernism; and the use of a proliferation of symbols that are held together precisely because the idea—in the dissolution of a synthetic and thus mythic unity—involves complexity and diversity.[10] However, prior to noting what occurs within this division, it is essential to realize that the predominant response to this position—a response that had its origins in a certain conception of modernism—was to opt for an apparently neutral style.

Neutrality was the identification of style that transcended boundaries. Once the idea of national identity as determining style gave way, one of the forms allowed by this collapse was a so-called internationalization of style. In fact, what occurs is something slightly different. The internationalization of style does not means that style has become international; it is rather that capital which is "neutral" has become international, bringing with it a ubiquity of style. This has the liberating effect of overcoming recourse to real or potential national utopias, since it is already the utopia of the international. What has to be noted here is that once the growing recognition of the impossibility of this latter form of utopia takes place in terms of projected national or ethnic utopias, then this runs the risk of defining opposition in terms of terror. The terror of the international—which is, of course, not international as such—cannot be met with terror of the national. Even an aesthetic response to the ubiquity of international style—other versions of the same—brings with it a culture defined by the opposition of boredom and novelty. One interrupts the other. Boredom engenders distraction. Distraction in wearing off yields boredom. One works with the other. This conflict would be merely aesthetic if the internationalization of style were simply that; namely, no more than a unified aesthetic form devoid of economic and political implications. The fact that it is not means that its symbolic dimension also has a ubiquity; one that will be bound up with the consequences of the internationalization of capital and the identification of that movement with certain nation-states or economic groupings.

The critique of international style therefore has to work on the one hand with the feint of neutrality, while on the other, it has to recognize that the nostalgia of the heritage industry or the realizable impossibility of arguing for the return of vernacular architecture—given that such a claim brings with it a conception of place and settlement that the impossibility of the national unity has already unsettled—are no longer options. This is the point at which it becomes possible to begin to distinguish between a number of different moments, to begin to see how, given this setting, the question of style can be addressed. The presence of differences reenacts the twofold move that has already been noted. Namely, they have the conditions of possibility in the severance of style and appearance. What they evidence, therefore, is not a simple proliferation of possibilities but of the advent of the modern.

Symbols

A proliferation of symbols without project would be the most apposite way of describing what happens when symbols are let free without a corresponding unified idea. (This is, of course, only the veneer of freedom: the one defined negatively by the absence of constraint.) What occurs is the move from the unity of the idea, in which the decorative did not play a role, to the use of symbols as simply, or merely, decorative. The decorative occurs once the unifying idea has vanished. However, to the extent that a unifying idea is displaced—and it will always have been the case that such conceptions of an idea, working through symbols, tried to express a singular vision, be it mythic or projective—the question then emerges concerning the possibility of the retention of symbols and thus of expression. It may be that the use of symbols after such a hold has been relaxed—relaxed because futile—could only ever be decorative. Hence architecture, if it were to use symbols, would do so only on the condition that they were ornaments without reference. (Not a deferred reference, but without a direct reference and therefore as inextricably bound up with the ornamental.)

Is this the answer to the question of style? If postmodern architecture is not post modern in any sense that bears a real relation to the functional aspect of modernism, but only to the question of function's appearance, then the postmodern has to be reconceived.[11] Rather than involving a temporal category, it becomes a move in the history of ornament; a move in which there is what could be described as the "ornamentalization of ornament." The direct result of this recategorization is that there will need to be another way of judging symbols. Furthermore, it opens up the possibility of symbolic dimension within contemporary architecture. The building that will be considered in this light is the National Museum of Australia (NMA) by the firm of Ashton Raggatt MacDougall (ARM).[12]

In terms of program, the NMA already has a defined relation to the question of identity insofar as it pertains to the question of national identity. In the case of Plečnik, the answer to the question of national identity was unequivocal. The failure of that idea and thus the necessity to recast the question of the nation within modernity becomes an urgent one. How is a national museum to engage with the question of national identity once the answer is no longer determined in advance in terms of a form of already given unity? The engagement has to have an architectural resolution. It must inhere in some sense in the building's structural presence. If national identity is a site of negotiation—and if this is a site that must be maintained as the locus of negotiation without end, precisely because there is structural equivocation in the question of identity itself within modernity—then once again, there has to be an architectural correlate to that state of affairs.

With the NMA the impossibility of the unifying idea—i.e., an already determined and synthetic conception of Australian identity—is met by the proliferation of symbols. The important point that has to be noted in this context is that simply because there is no longer a unifying idea, this does not mean that there cannot be symbols; though as will be suggested, while they are symbols they fall beyond the conventions of the symbol. If the convention demands an immediate move between the symbol and the symbolized, then what defines the symbols on the NMA is an absence of immediacy. If they are symbols, then they work in fundamentally different ways from either postmodern symbols or symbols defined in terms of immediacy. (It is immediacy that allows the symbol to be unifying while at the same time existing as a set of identification.)

The argument that needs to be advanced has to start with the position already noted; namely, that the NMA is not defined by the postmodern retention and use of the symbol as the merely decorative. Postmodernist architecture deploys symbols in a very specific way. They become the sign for the absence of a singular project, and that absence has to be understood as the mark of a resignation, even if it is a resignation that redefines the architectural in terms of its capacity to please and thus to decorate. In contradistinction to a conception of symbols that appeals to no more than a generalized subject of pleasure such that architecture comes to be dissipated within a purely aesthetic realm, there can be the retention of the symbol. Postmodernism operates by deploying dead symbols. (For Plečnik symbols had life.) This retention still operates in relation to symbols. There is another possibility for the symbol. The strategy of the NMA's use of symbols needs to be situated in relation to this other possibility. Rather than an overarching idea that organizes the symbols, their presence is dispersed across a field, thereby constructing a dispersed field. In addition, there is an internal dispersal within the symbol itself, insofar as each symbol carries more than a given and thus immediately accessible conception of the symbolized. Immediacy, understood as temporal simultaneity, inscribes a different temporal status with real spatial effects. In addition, the overdetermined symbol renders the return to immediacy impossible. In the opening, a spacing that marks the temporality of the continuity of discontinuous negotiation, immediacy yields its place to the mediate. What allows for mediation is, of course, time.

At the National Museum, identity becomes a site of endless negotiation and the symbols work to maintain that position. The symbols—their particularity—become the architectural resolution to the generalized question of national identity. Symbols are as such overdetermined—Le Corbusier's Villa Savoye becomes black. Such a move involves a negotiation with modernism—its whiteness—as well as the question of color. The negotiation is carried by the now overdetermined symbol. However, rather than concentrate on the symbols per se, what is actually fundamental to their presence is that they introduce a conception of time that is not determined by immediacy. There is a position, as has been suggested, that links the symbol to a more complex conception of identity. Complexity in this context does not mean that identity is undetermined. Rather, it means that it is always to be determined. Identity names the process of negotiation over its own nature. There is still the link between symbols and symbolized; however, what needs to be noted is that the link is hard to establish. The comparison with Plečnik could not be more dramatic.

While it is possible to extend this argument, in the first instance, by analyzing in detail the symbols that make up the NMA, and in the second, by showing how the initial geometry generating the design (the adapted Boolean knot) is integral to any account of the building's operation, the most significant element for these present concerns is that in the move from immediacy to mediacy, there is the inscription of spacing at the heart of the building's own spacing. That inscription has two consequences. Both need to be noted. The first is that the consumption of the building and therefore its capacity to offer a site of engagement—and perhaps also to give pleasure, pleasure as architectural effect—occurs over time, and that with this time a site of both criticism and reflection occurs. In other words, the symbols in not being readily consumed and thus by eschewing immediacy yield a spacing that has time at its point of inception.

The second consequence involves site. The argument in this context is utterly specific. It is in terms of this opening that site as a neutral territory—site as an architectural tabula rasa—is called into question. However, it is not any site. Australia becomes the place. Now, however, it is Australia as land marked by an ineliminable conflict between an indigenous population and settlers, a conflict demanding recognition, and legal entitlement at the minimum. And even then the combination is more complex. The relationship to land changes over generations. The influx of immigrants transforms neighborhoods, leaving earlier residencies and populations that were there as mere traces, vestiges, or only ever to be recalled in acts of memory. All of these elements can be construed as lines that cross and intersect. To the extent that this occurs, the literal site on which the lines cross and knot becomes complex. The overall layout of the ground plan of the museum building, as well as the "gardens" that form part of the complex, dramatize this exact possibility.

In sum, therefore, the importance of the NMA by Ashton Raggatt MacDougall is found in the way the symbol breaks the hold of immediacy. The introduction of another temporal possibility—one having spatial effects—means that this "break" is inherent to the architecture and that as a consequence, the complex as a whole needs to be understood as one possible architectural resolution to the exigencies of the affirmation of modernity. The limit of the project resides in the extent to which overdetermined symbols can be deployed. Given the severance of style and appearance, the identification of this as a limit, coupled to the openings that the identification of this limit secures, are both internal to the project of cosmopolitan architecture. As such, they indicate both the position and the effect of architectural autonomy at the present.

Surfaces

As has already been suggested, two related conditions mark the inevitability of the problem of decoration within contemporary architecture. The first is the demise of the guiding single idea. The second is the impossibility of attributing to architecture the instrumentality demanded by such an idea. ARM's response was to use the symbol, or rather to reposition it within a structure of mediacy. Another response—and it should be remembered that these responses have similar conditions of possibility, which accounts as much for their difference as it does their connectedness—is to begin to define the architectural in terms of the surface. Rather than a productive sense of surface, the first instance in which such a positioning operates is in terms of what will be called the "literal surface."

In order to develop this point—a point resisting the arbitrary precisely because of its playing an evidential role in relation to the modern—two projects will be considered. The first is Herzog and de Meuron's Eberswalde Library in Eberswalde, Germany, and the second is the Online Multimedia Centre at the St. Albans campus of Victoria University in Melbourne, Australia, by Lyons.[13] Both buildings are the consequence of contemporary production techniques. In the case of the Online Multimedia Centre, the surface is constructed out of a compressed sheet cladding system using urethane paint, while the production of the surface design involved CMYK digital processing. (The building is used to house computer laboratories.) In regard to Herzog and de Meuron's library for the Eberswalde University—the building was commissioned in 1994 and built in the years 1997 to 1999—it was undertaken in conjunction with the artist Thomas Ruff, who worked together with the architects on the design of the facade. The building brought together two motifs that dominate recent work by Herzog and de Meuron, namely, the box and techniques of complex facade construction.[14] While as a motif this use of the facade gestures toward the use of sgraffito in traditional Swiss architecture, Semper's interest in textiles, and Otto Wagner's ornamented facades in Vienna, a great deal more is at stake in the way the consequences of this technique unfold.

Of the many elements that unite these two projects, perhaps the most complex has to do with the buildings' relation to time. On one level these buildings are images. And yet they are images in a very specific sense. Their isolation and analysis—as images—involves a process in which they are stilled and then discussed in terms of the relationship between the techniques of surface construction—and it is always the construction of literal surfaces—and the relation of those surfaces to function. Within this setting, it becomes clear that while Lyons's building allows for a reworking of a given programmatic constraint, namely a computer laboratory, in that the surface allows for the presentation of a colored facade while maintaining at the same time a consistent temperature necessary for a computer laboratory, once the images are stilled the nature of the function becomes irrelevant. This is clear from an analysis of the plan, though equally from the location of the architecture in the surface. However, to insist that the surface and the image are the only sites of architectural analysis would be to misunderstand the nature of the images in this context. The conjecture is that they are images in the very strict sense that a frame is an image from a film. Once a film is projected, then the images are concentrated within and therefore integrated into a field of continuity. Not only does this provide a way of approaching the implicit and explicit urbanism of these buildings—the implicit is the building's inherent urbanism, the explicit the way that urban presence may be an intentional part of the building's structure—but it also allows the nature of the surface to be addressed. As an image, it oscillates between the time of the frame and the continuity of a field of operation that involves virtual presence as much as it does actual presence. Indeed, it is possible to argue that the two would coincide. This latter sense of time is not narrative time. On the contrary, it is that conception of temporality that allows for an account to be given of the insertion of an image into a domain of activity where that setting holds the image in place while dispersing it. The dispersion is the virtual field of images: what can be described in this context as an urbanism of surfaces.

Their urbanism—and it must be emphasized that it is as much implicit as explicit—means that the time of their consumption is, in the first instance, the temporality of the instant. The singularity of the moment in which the frame is frozen and the relationship between surface and function that is captured by the plan can be shown is one form of the instant. The vanishing of the particularity of the frame into the play of surfaces that becomes the place of the image within the larger field of activity means that the instant has given way to immediacy. What was there as an instant of pure givenness now enacts the insertion and the vanishing that mark the dispersal of an image within the potentially endless becoming of images that marks the presence of a film. While both these buildings stand out in relation to their context, the fact that they also have the potential of redefining that context allows for them to be reabsorbed. An architecture of literal surfaces generates sites—instances within continuity—that have to be addressed in terms of speed. No longer is it the same as the immediacy of the symbols/symbolized relation. Here another mode of thinking can emerge. It is one that has to privilege absorption and individuation. The architecture—and it is the architecture—moves between its presence as an image to the movement of that image, its movement as a frame, within the urban film. One of the consequences of an architecture of speed is how a concern with function would be introduced. That Herzog and de Meuron do not address it results from the way they privilege the box as the conveyor of the surface as image. Whether this is the only possibility for such an architecture remains to be seen. It must be added that allowing for the complex temporality of the image opens up the question of place. While there can be no satisfactory response—as yet—to the question of how this place is to be understood, what is important is that it positions such questions beyond the normalizing—perhaps also moralizing—language of compatibility between environments. In other words, emphasizing speed and thus a visual urbanism undoes a conception of the urbanism driven by visual contextualism. It works in this way since contextualism addresses the presence of the surface as a single image determined by the temporality of the instant. Speed positions the image within the generality of dispersion.

And yet a further caveat needs to be added. There will always be a need to distinguish between two senses of visual urbanism. In the first instance there is a visual urbanism that involves the centrality of the brand. The latter is no more than the urban presence of a universal sign system. Again, the universal in question has the same status as the "international" character of the so-called international style, namely, the reiteration of signs that are the branding of international capital. In the second instance, however, there is the possibility of a presentation that, in resisting the determinations of both the national and the international, takes on and thus displays another form. The mingling of both possibilities positions any visual urbanism—now understood as a more generalized field—as a site of judgment.

Allowing for visual urbanism—in the complex sense—introduces a further element that works to redefine place. This does not occur because some inherent truth about place is revealed. Lines of differing speed and intensity, objects articulated within forming part of them. Lines carrying traditions of conflict, agreement, and negotiation proved the truth of place. In addition, emphasizing speed allows for the ineliminable presence of the urban as the locus of human sociality to be noted. While terms such as the "urban" and the "city" may bring with them as many complications as they provide potential for clarification and more detailed analyses, understanding the interplay of the question of style and the cosmopolitan—perhaps the site of style within modernity—is a necessity in order to understand the space opened by the forced abeyance of an architecture driven by the symbolism of immediacy. Surfaces creating part of a visual urbanism—whether potentially or actually—need to be considered in that light. The interplay of spacing and distancing demand that time play a pivotal role in any account of the architectural.

Occurring with the symbol—as conventionally understood—is a different temporal order. Convention demands that the symbol be a locus of immediate identification and comprehension. However, once the symbol becomes complex such that it engenders a locus of negotiation—as was instanced above in the discussion of ARM's National Museum of Australia—where questions of meaning endure as questions rather than as yielding automatic responses, then what is introduced is another temporal dimension. The temporality of "mediacy"—of discussion and hence dissensus—introduces a twofold movement. What structures that movement is the interplay—and it is a necessary interplay in that one needs and bolsters the other—of difficulty and democracy. ARM's National Museum locates the democratic impulse in the building's symbolic presentation, What will be argued with LAB's Federation Square project is that the democratic is located in the opening of an urban field of negotiation—a place of meeting and dispersal. Again, movement and through movement time come to take a defining role. What is distanced with this project—and the inherent critical dimension of this distancing needs to be noted—is the tradition of the urban square and its inherent monumentality. It should be added that this is what the architecture effects. It is not just effect. It is the project's work. It is only in the formulation of the presentation of an object structure as a site of resolved irresolution and thus a site of dissensus that the work attains autonomy. Yet because it is architecture and not art, autonomy is inextricably connected to modalities of use. This has an important impact on how site and place are to be understood. The democratic necessitates criticism. This will have the interesting effect of the refusal of the space of novelty. Novelty is not simply integral to the temporality of fashion. Novelty must be both immediately consumable and immediately disposable. The democratic is bound up with difficulty, and it is difficulty that stands in the way of novelty's necessary relation to immediacy. The spacing of difficulty and the timing of spaces beyond the immediate begin to mark out the cosmopolitan.

Urban Squares—Public Galleries

The Federation Square project, which officially opened in October 2002 in Melbourne, is easily the most adventurous urban and architectural program that has ever been undertaken in a major Australian city. The LAB Architectural Studio (which worked in conjunction with Bates Smart) operated on a site of 3.6 hectares (8.9 acres). Not only was a new urban square created, but that creation included four major public buildings. Rather than try to do justice to the whole project, in this instance emphasis will be given to the relationship between the urbanism of Melbourne and the internal organization of the National Gallery of Victoria (the Ian Potter Centre), or NGV. While it is always problematic to choose one element of a project, what justifies it in this instance is that the project as a whole needs to be accounted for in terms of the centrality of the urban. Moreover, it is an urbanism that does not reduce the buildings to figures that occur in an urban ground. What the project as a whole demands is a reconfiguration of the urban condition of Melbourne's Central Business District (CBD), in order that the oppositions of figure/ground or building/context can take on a different quality. Rather than privilege any one element of these oppositions, centrality has to be given to a series of relations occurring within a general condition. What occurs is a process that individuates. In more general terms, repositioning the urban context in this way not only underscores the importance of the urban as the condition of human sociality; it also allows for a presentation of the urban in cosmopolitan terms. Again, this is the argument.

Once primacy is given to the public nature of architecture, such that a building's implicit and explicit urbanism is given centrality—a centrality that turns the house into the exception rather than the norm—then the question that has to be addressed is how architecture's complex urban presence is to be explicated. On one level, the presence of architecture within the public domain might be thought to account for its urbanism. And yet that is not enough. All the identification such a level addresses is the urban cultural context. Admittedly, it is a context that is often denied, such that architecture is discussed as though the edge between the building and the context marked a real architectural division. (A division existing in itself rather than as a site to be explored.) Nonetheless, what this opens up as a possibility is that the urban may become a structuring force within the building. What this will mean is that the modes of occupation and the presence of a range of programmatic possibilities, both necessarily bound up with movement and therefore with circulation paths, will demand an account in terms of the urbanization of architecture. Such an account would allow for the development of another way of understanding the public nature of public architecture.

Caution is necessary here since the urban—as both a site and a condition of movement—does not form an ideal type. Public architecture always allowed for circulation. Circulation did not simply replicate social conditions. However, circulation assumed important forms of distinction. While distinctions remain—it is not as though issues concerning safety are untouched by questions of race and gender, for example—it is also the case that the contemporary urban condition has to be defined in terms of movement through time. (What changes, in fact, is the conception of the public; the public has a history.) Space, currently, is a condition that can take movement and time as that which begins to determine its parameters. These considerations need to be located within the setting that most typically can be taken as defining the urban condition; namely, the grid. Once the question of the urbanization of architecture is allowed—a concession that acknowledges the structuring potential of the urban within architecture—then what has to be addressed is continuity. How does the urban condition continue?

The hold of the grid works by allowing for two fundamentally different forms of continuity. The first is a type of reiteration where the structuring force of the grid is repeated with an almost visual form of continuity. Here the grid can proliferate ad infinitum. Moreover, the grid can continue vertically as well as horizontally. (In other words, the logic of the grid can be present as much in section as it can in plan.) As such, the question of continuity is already resolved in advance. The other form of continuity is decomposition. While the initial presence of the grid sets the conditions allowing for its reiteration, it is the introduction of a logic that is discontinuous with the grid that will occasion another form of development. If the internal development of the grid is a repetition structured by sameness, then the intrusion of another logic allows for a reiteration structured by difference. Recomposition occurs because the grid is continually traversed by movement. It is, of course, a movement, at least initially, that the grid allows.

What is of interest for this project is this second form of continuity. And yet, even in accepting the centrality of movement, movement on its own is not sufficient. Flows are not enough. Movement is always accompanied by arrest. Within the grid, movement works in relation to attractors. These attractors—their location within movement, *within* and therefore as part of a generalized process—allow for further movement. Recomposition—within the grid—works in relation to the complex interplay of movement and arrest. Complexity arises because both these terms differ in intensity through time. (It may also be that differing intensities mark a form of temporality other than the strictly chronological; in other words, the temporality proper to complex subjectivities.) In general terms, once time is introduced, then what vanishes is any possibility of seeing the relation between movement and arrest as constant. This inconstancy, however, is not simple chaos. What precludes chaos is the body. The urban body works on at least two levels. The body moves through spaces. The movement is defined by positions. Moving through positions. Equally, the body sees. Movement may be defined, therefore, by what can be seen. Objects of sight are coterminous with bodily presence. At the same time, the body sees positions—and hence positioned objects—that are not coterminous with bodily presence. Indeed, what the body sees may be what cannot be reached. Seeing and positioning, their conjunctive and disjunctive relation, define the body that moves. Equally, it defines the body that is continually located and relocated by attractors that structure the relation between movement and arrest.

While this description is almost of necessity quite general, what it sets in play are the conditions for understanding a form of continuity that is defined as recomposition rather than the simple continuity of sameness. Even though the detail of this position stands in need of greater adumbration, part of what marks out its importance is the way it enacts a division between form and appearance. If the continuity of the urban, where the ground to be made continuous is the urban grid, comes to be defined as recomposition and where recomposition has to integrate the complex relation between movement, arrest, and the body, then this will allow for an instantiation of the urban which, while having no visual relation to the structure of the grid, can nonetheless be defined in terms of continuity. This time, however, it is continuity as recomposition. In other words, what this allows for is an account of the urbanization of a given domain that is not simply the extension of a preexisting structural condition. This is especially the case if that condition is understood as the literal grid.

Historically, the initial grid system defining the CBD of Melbourne began to open up due to the intrusion of lanes causing divisions within the overall structure. While this opening up is internal to the system, urbanism occurs as a consequence of movement through the streets and lanes that identify the area. In other words, urbanism is not just the result of a reiteration or alteration of an existing pattern. Movement and arrest are integral to the process of urbanization, and hence it will always have to be more than a structural concern. They involve additional considerations. In this specific example, the presence of transport hubs and other lines of dispersion (notably an arcade system) occasion movement through the grid. While the hubs and lines are identifiable within the structure, they are not a necessary part of it. Movement would remain a single flow if the occurrence within the grid of different pathways or lanes did not allow for the emergence of program as attractors. (Attractors arresting movement and therefore opening up other programmatic possibilities.) In addition to the existing structure of arcades that provide a passage through the grid and which function as attractors in their own right, the presence of lanes and streets indicates the extent to which the existing system takes on different qualities—and thus is recomposed—by the interplay of movement and arrest. Once this interplay is allowed to form an integral part of the definition of the urban, then it becomes what is replicated in an urbanization of architecture, the latter being a process of inscribing a string conception of the public into public architecture.

The early composition studies for the project refuse any attempt to rid either the setting or the context of an inscribed urban presence. At no point did context stand opposed to the figures which would become either the National Gallery of Victoria (the Ian Potter Centre) or the Australian Centre for the Moving Image. The studies bring lines into the site. The lines begin to allow for individuations that open up the possibility of an architectural intervention. The move from compositional study to architecture involved a number of factors. In general terms, what can be said to have occurred was the creation of a public square in which different institutions could be given a place without reducing their presence to figures within a given terrain. The site is taken as a field. What constructs that field condition involves pulling lines of movement through it. It is in relation to this process that volumes begin to emerge. They emerge not as additions but as individuations within a generalized terrain. While the terrain is generalizable, it needs to be emphasized that external elements, movement, etc., would be central to any account of the individuations. It should be remembered that the process of individuation answers the question of how the relationships between volume and context emerge.

The urbanization of the National Gallery of Victoria—and here what is involved is the urbanization of architecture—lies initially in the way lines of sight occur within the building. The lines occur in the plan as well as in the elevations. On the first level of the building, while the galleries can be seen as traditional boxes with an interstitial space allowing for entry and thus movement from one floor to another, such a definition does not begin to address the question of their internal complexity. Again, it is not the complexity of chaos. The urban body intrudes. Hence, it is possible to argue that organization within the gallery spaces—the body's organization and thus its realization of program—has to do with different lines of sight. In the galleries devoted to nineteenth-century Australian art, for example, it is possible to redraw the internal relations in at least two different ways. (These two possibilities are represented in figures 1A and 1B. The former is the actual floor plan with lines of sight sketched, and the latter is the same floor plan emphasizing these lines of sight.) The first way—figure 1A—takes the internal divisions as given. The divisions therefore become simply internal spaces within a box. Equally, however, it is possible to draw a line from the northern corner of the top gallery that cuts through on a diagonal to the right southern corner.

Once this is done, another line of sight emerges that moves from that corner through the box. The process can be repeated throughout the boxes comprising the galleries. And yet it is not as though there are two different ordering systems at work. The lines of sight are not additional elements. Once taken in conjunction with the divisions within the gallery spaces, they mark the presence of the conception of sight (and thus bodily presence) that can be taken to characterize the urban. This process can be identified on the ground level as well as on level 3 of the building. There are many studies of the lines of sight in the documentation accompanying the building. While they cannot be subjected to a single interpretation, what they indicate is that the urban, once understood as the relationship between movement and arrest located within an existing structure, can be taken as the generator of the initial design process. While it cannot be detailed here, this process occurs equally in the sections and accounts for the nature of the relationship between the intrafilament and the gallery spaces.

The overall effect is that program is given a more complex determination. The question of the display of art is opened up. The opening is caused by allowing the urban body to play a significant role in the generation of form. While the presence of the urban body may not have been the actual prompt, the engagement with the display of art is moved away from the white cube. No longer is the viewing of art structured by the body's relation to art being defined by the floor, the corner, and the wall. In other words, the move from the white cube occurs because the conception of the body demanded by the cube—which is the solitary body in private space—is no longer in play. (Pursuing this point would become a way of arguing that a certain conception of the viewing body has a critical dimension.)

This assessment can be taken further, since it can be argued that the relationship between lines of sight and the internal divisions of the gallery spaces working together define programmatic activity. What cannot be separated, therefore, are programmatic considerations and the internal geometry of the building. They are tied together. (Again, this position is as true of the plans as it would be of the sections.) Part of the significance of the NGV is that complex geometries are interarticulated with programmatic concerns. Here there needs to be a real distinction drawn between the relationship between program and design in the NGV and other museums where

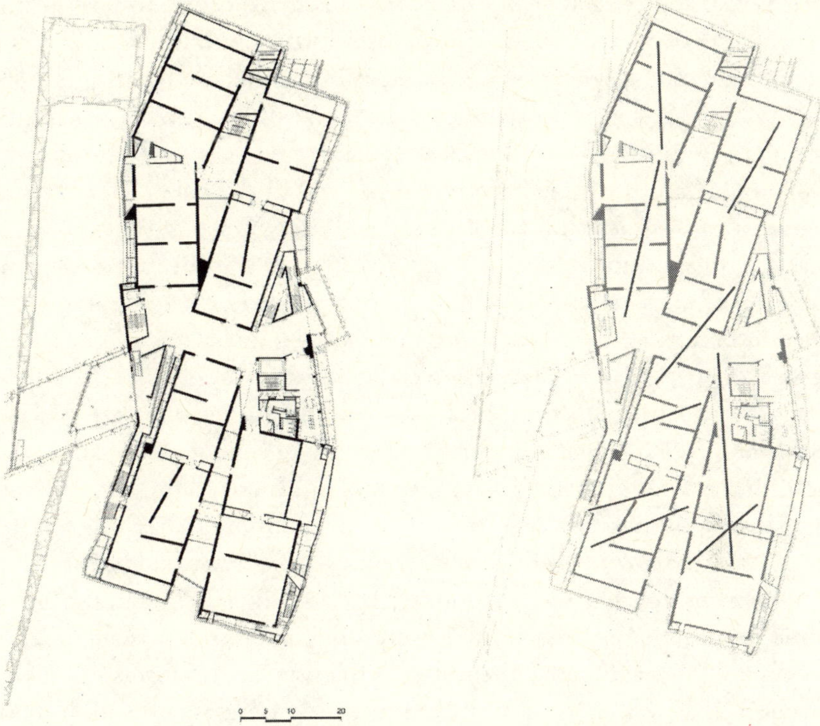

Figure 1A **Figure 1B**

Plans of level 2 of the National Gallery of Victoria at Federation Square

the use of complex geometries remains, for the most part, indifferent to programmatic concerns. Innovations, in such instances, are literally superficial. They endure on the surface leaving a more nuanced sense of architecture untouched. Complexity becomes no more than decoration. While the decorative can be linked to a building's acquisition of an iconic status—the so-called signature building—if programmatic concerns are indifferent to its external organization, then this severance inscribes architecture within a conception of the spectacle that will only ever enjoin complacency.

The National Gallery of Victoria will have to play a significant role in any future discussion of the relationship between the facade and the internal—hence programmatic—organization of buildings. While it is true that the internal organization cannot be read from the exterior, and thus there is a structural indifference between the exterior and the interior, the presence of the fractural facade cannot be understood in terms of the decorative versus the functional. Of the many aspects of the building that preclude such a reductive approach, perhaps the most important in this context is that such an approach would insist on seeing the building as a figure located in a ground, where the latter is architecturally distinct from the building itself. What occurs in Federation Square is, as has been intimated, the opposite. The site itself is a constructed field in which volumes (buildings) become individuated elements within the field. To the extent that such an argument is maintained, then the fractural facade begins to repeat the organizational logic and hence the geometry governing the construction of the square itself. While there is on one level an important distinction between an exterior wall (here a facade) and a square, once they both form part of the same domain of activity—the field—then what emerges as a consequence, in this instance, is that the relations between the facade and the square are of greater architectural significance than the relationship between the facade and the building's interior. In sum, it is simply naive to think that architecture concerns no more than a relation between an external wall and a structured interior. What would be forgotten in such an approach is the position and effect of the building in the urban fabric. In this context, this is not simply a point about the visual presence of the buildings and hence their visual urbanism; it also forms part of an approach that refuses pre-given distinctions between figure and ground.

The urbanism that works to construct the interior of the National Gallery of Victoria is also evident in the construction of Federation Square. However, precisely because there are different programmatic constraints in each instance, the appearance of the urban as a generator of form will be markedly different. As has been indicated, privileging the facade/interior relation fails to allow for the role of the urban within architecture. Or, more exactly, it reduces urbanism to a given. The continuity of the urban becomes no more than the literal extension of a preexisting structure. What is literally there is repeated literally. What Federation Square opens up—and this has been the argument concerning the urbanization of architecture as evidenced by the NGV—is a different and more productive way of understanding the continuity of the urban and the position of architecture within that movement.

What emerges is a site that can continue to be used, since the conditions of use would only be gradually acquired and thus never fully mastered. The convention of the square with its inherent monumentality gives way to the urban square. The monumental and the given cede their place to the difficult and the democratic. "Difficulty" is not just a term that concerns understanding and thus an attempt to attribute meaning. If this were all that was involved, then architecture would be the modernist poem. In architecture, difficulty does not pertain to hermeneutics; it does not have to do with the complexity of coming to hear. Difficulty in architecture involves time in at least two interrelated senses. In the first sense, it is the time of the democratic, the endless negotiation over that which is not given immediately. The second sense is bound up with the first, since now time becomes the problem of use; perhaps more accurately, use as a problem. Time here involves learning to use a building, and thus there is a new constraint; namely, of having to find oneself and others within an urban space. Of coming to see potentials, of working possibilities, of opening up terrains that hitherto had been closed within the confines of the tradition of the urban square with its flat side, abrupt corners, and level surface.

Here place emerges, since the complexities of time are placed. They are, however, not resolvable by reference to the literal place. Once this can be allowed, then the presence of concrete determinations will always be that which is worked through. In the opening, in the refusal of finality—though with the necessity of ends and means—in the possibilities allowed by speed, there is the harbored presence—more accurately the site—of a cosmopolitan architecture.

Styles of Building

In moving from Plečnik through to the concerns of LAB, a particular path has been traversed. In the first instance, the argument is that while the move from an idea which holds the architectural project in place—Plečnik's "inner nerve of art"—to the differing possibilities occurring as a consequence of that move seems to have taken on the form of either relativism or a type of architectural pluralism, this is not the case. What sustains the argument is that if it is right to argue that Plečnik is the last moment of symbolic architecture where the symbol is marked by immediacy, then this needs to be understood as another opening up of the question of style. The question is doubled. Not only is there the severance that has already been noted between style and appearance. At the same time, there is a break in the relationship between architecture's symbolic dimension and the organizing structure of the idea. This is not to suggest that buildings don't have symbolic presence. Rather, the symbols become situated within conflict. Overall, therefore, the cosmopolitan—in architectural terms—has to be situated within the severance of style and appearance and as such becomes the possibility of an architectural practice which, while refusing a conception of final form that is always determined in advance, equally has the possibility of overcoming the hold of novelty. The question of its appearance is therefore central. What appears needs to be addressed in terms of its particularity. There are conditions of judgment, and yet those conditions are not grounded in a converging notion of generality or an all-encompassing sense of universality. The absence of generality, as has already been noted, makes the question of judgment inextricably bound up with both the advent of modernity and the potentiality of the cosmopolitan.

6. Refugees, Cosmopolitanism, and the Place of Citizenship

Two demands accompany the refugee.[1] The first demand is a political one. There has to be a response to the situation in which an apparently sovereign state is beset by those who wish to live in it. Those without citizenship—or who have a citizenship that places them in danger—seek refuge and perhaps, in the end, the citizenship of a separate country. The displaced seek a place. They land as refugees. The complicating factor emerges at this point. Despite the insistence of a practical political problem, even in formulating this positioning of the refugee, the language of classical political philosophy with all its apparent neutrality is simply assumed. In addition, what is presupposed is a certain geographical understanding. It is as though terms such as "place," "land," and therefore "displaced" and "seeker of refuge" are all understood. The problem is that their understanding is given by, and of course within, the language of a political philosophy that the presence of the refugee may not simply jeopardize, but may cause the necessity for a fundamental revision. This is, of course, the other demand. The point is straightforward. The refugee, the position, perhaps even the humanity of the refugee, demand a response. And yet the dilemma is that what the presence of the refugee causes is a crisis in the conceptual framework, and in the language and terminology of that framework, in which it is assumed that a response would be given. This second demand involves the discovery of a language and an interrelated mode of thinking.

In part, the difficulty in formulating any opposition to current policy in relation to refugees—given that there is no opposition from the Opposition—is finding a language and conceptual framework within which to articulate another view.[2] The project of this chapter is to begin to take up that task. What needs to occur is a fundamental shift in which the presence of the refugee is not viewed as an exception. The problem of the exception is that it generates a simple either/or. Either all refugees are excluded in the name of homogeneity. Or the refugees that are admitted are allowed entry on the condition of assimilation. (Exceptional status is either maintained or overcome.) The link between entry and assimilation is also undertaken in the name of homogeneity. Even if the homogeneity is not thought in cultural terms—and it should be added that it often is—it is present in legal terms. In this context, law is of fundamental importance. The legal framework has long assumed a type of neutrality, and as such maintains a refusal to allow itself to be the site of different and at times incompatible legal systems. In the Australian context, the Mabo decision has made that state of affairs a real possibility, one where a potential dissensus would be the ground for a future consensus. Once it can be argued that the extension of Aboriginal law as it pertained prior to settlement is greater than property rights—in other words, that even though Mabo was restricted to property rights, there is nothing inherent in the argumentation to limit it in this way—then there is in principle no objection to extending customary law to all other areas of custom and tradition. The consequence of such an extension is that the category "subject of right" has to be understood as admitting an ineliminable complexity. There will be no one subject of right, since that which determines right—law—will itself be the site of different possibilities. In this context, it is vital to note the comment made on a recent Canadian case—one mirroring aspects of the Mabo judgment—in which part of the court's decision involved the argument that

> the only fair and just reconciliation is one which takes into account the aboriginal perspective while at the same time taking into account the perspective of common law. True reconciliation, will, equally, place weight on each.[3]

What is significant about this argument is that it necessitates a conceptual shift in which the tenets of conventional liberalism are no longer applicable; what has to be discovered is a language and terminology able to articulate the results of such a position. It will, for example, no longer be possible to hold to a straightforward distinction between law and justice, a distinction which envisages their coincidence but which cannot guarantee it. The contention here is that the challenge posed by the Mabo decision, once given greater extension, also occurs with the refugee. In order to address the insistent reality of the refugee, there has to be a shift in thinking. Moreover, that shift in which terms will need to be rethought, will in the end be bound up with precisely the concerns for law and the subject of right that flow from the consequences of allowing Aboriginal law extension while at the same time retaining the place of common law. In the same way as the potential for reconciliation was there from the start, the presence of the refugee—once the terminology announcing its presence is itself rethought—is not a recent invention.

This is not to argue that the position of the other or the stranger has a history. Such a claim is clearly true. Rather, the point is a different one. The claim is that once it can be accepted that Aboriginal law had an existent reality, in the same way as there is a conception of place that forms an Aboriginal connection to the land, then any subsequent relation between them and later conceptions of law or place opens up the possibility for reconciliation. Equally, of course, it allows for the denial of reconciliation. What was always there was a potential, even if that potential was refused. The refusal of an already present potential and then subsequent arguments and policies designed to activate that potential give rise to a different set of philosophical and political conditions than those that demand and enjoin the granting of rights where they were thought not to have been present.

Positioning the Refugee

The problem of the refugee and the problems posed by refugees endure and have endured from the start. Once the city wall as an empirical entity, or a conception of land that necessitates a territory that may be owned or possessed, are taken as original conditions, then movement across a border or through a gate is inextricably bound up with human existence.[4] This is not a claim about property per se, but about situatedness and therefore about place. What this means, of course, is that from the start the question of the refugee—understood as the movement of people from one possessed state to another, a movement originating in or resulting in dispossession—cannot be radically disassociated from the history of exploration. (Part of the argument to be developed in this context works with the assumption that once movement is central and that therefore once settlement has to be understood as the arresting of movement—occasioning further movement—then rather than the refugee being the exception, it becomes a form taken by movement.)[5] Why should it be assumed, therefore, that the refugees are the weak—since what is at work is an instance of subject positions defined by movement—and that therefore the problem of their reception has to be one in which a sovereign power has to negotiate its sovereignty with an uninvited guest? Why could it not be the case that the refugees—though now the term will be stripped of its pathos and be defined by movement—are strong and well-armed, and that a sovereign and indigenous people are caught within the dilemma of reception and hospitality? To be hospitable is not a question when an invasion takes place. If the history of European exploration and land acquisition is the extension into another realm of certain presuppositions about the nature of citizenship and its ineliminable relation to the "ownership" of land, then the presence of the refugee will have been given a different setting.

What this means, though here argumentation is necessary, is that once movement defines place, then the refugee, rather than being the exception, actually defines the setting of place. Movement across and through places—though the quality of that movement and its relation to borders and laws will always be changing—allows for settlement to be reinterpreted and thus an original condition between citizenship, right, and place to be questioned. Questioning does not mean abandoning. Rather, instead of beginning to define situatedness in terms of a founding relation to place—and such a relation quickly allows a conception of place to be interarticulated with ethnicity, and hence "blood and soil"—what should be done is to see settlement and finite conditions as the momentary arresting of movement. Those moments can be mythologized in terms of national histories. However, the demythologizing move is always one that shows that discreet moments—moments of fixity and settlement—are defined and made possible by movement. Movement will remain as a potential integral to any definition of human being. In these terms, therefore, the city wall and the establishing of borders become necessary conditions insofar as they establish what is fundamental for all conceptions of human sociality.

Argued for in this way, what is then introduced is a tainting of what was thought to be pure. A tainting that emerges as original. A conception of the citizen, an understanding of sovereignty, and their articulation within law—whether that law is common or statute-based—are not abstract positions that can be divorced from their role in the structuring of history. They are documents of culture. In 1940, as the crisis of European identity had already assumed stark proportions, and when the resources of a conception of general citizenship so dear to the Enlightenment were in a state of collapse, and their extension into constructions of national citizenship were being used to deny right by denying citizenship, Walter Benjamin wrote with great insight that "there is no document of culture that is not at the same time also one of barbarism."[6] Benjamin's identification of "barbarism" is the recognition that within the neutrality of a terminology that is thought to define objective legal categories, something else is involved. That additional quality is not an extra meaning; a semantic overdetermination. More is involved than the semantic instability of terms. What is there in addition—though it is an addition that defines objective criteria and thus should cause their redefinition—is what occurs when objective criteria are given the force of nature. Nature, however, would be that which had the quality of the immutable—a temporality that can only be secured within the structure of mythology. The interplay between settlement and movement—settlement as the condition in which movement ceases—comes to define human life. As will be seen, one of the important consequences of this position—and it will have an impact on how issues pertaining to welcoming are understood—is that to the extent to which movement and settlement are defined as fundamental, then what is important is not simply the way the refugee is met, but the framework within which the arrival takes place. If movement characterizes both the refugee who is in flight from his or her situation, and the movement of the explorer who lands and appropriates, then the difference between them concerns, at the beginning, the form of that arrival. If both are conditions of movement and thus to that extent both can be described as seekers of refuge, then what matters is the way the presence or absence of hospitality structures the arrival and not just the reception. What will need to emerge is the way that hospitality can dictate the way an arrival takes place as much as it can dictate the way that an arrival is received.

Viewed in more general terms, it should not be thought that the inherently problematic nature of the Enlightenment has not been demonstrated before.[7] What is important is not the problem as such, but the recognition that the question of the refugee is bound up with the interplay between citizenship and how the ownership of land is to be understood. Once this setting is allowed, then the issues raised by the refugee, and the response to those issues, cannot be clouded by simple moral propositions of the sort that it is essential to be kind, or that there is an obligation structured by a certain conception of charity. The opposite is the case. Once it can be asserted that the refugee is the problem of how identity is to be understood—insofar as citizenship cannot be easily separated from questions of identity—and in relation to how land and place are to be envisaged, then what emerges in the place of the homilies of morality are the more exacting determinations of ethics. Ethical thinking has its ground in ethos, and therefore in the complexities of life. What this means is neither life as a pragmatic dimension—though that will figure within it—nor life as an abstraction from the actuality of the everyday. Rather, it is life understood as the original human condition. The significance of life is not just its inescapability, but that it is inherently bound up with human sociality. For the human, life is sociality. This is, of course, Aristotle's insight when he defines the human in the *Politics* as a political animal. What is meant by that expression is not that humans are political animals; such a description leaves the question of the political untreated and the nature of the animal unresolved. What it means is that place is not just the locale where human activity may occur. Rather, place understood as the polis—and this is what place for humans means—is our belonging-together. This is human animality.[8] Belonging necessitates place. Place demands borders and frontiers. Once this is accepted, then the question of the refugee is inherent therefore to the very structure of sociality.

This is the context in which the relationship between the refugee as the sign of an original site of power and difference needs to be incorporated into a conception of human polity. While this is the conclusion drawn by a number of important commentators on the postcolonial condition, it is a potential that, in some sense at least, was already there. In the words of Homi Bhabha, for example, the postcolonial has resulted in a situation in which there is the need

> to rethink the profound limitations of a consensual and collusive "liberal" sense of cultural community. It insists that cultural and political identity are constructed through a process of alterity. The time of assimilating minorities to holistic and organic notions of cultural value has dramatically passed. The very language of cultural community needs to be rethought from the postcolonial perspective, in a move similar to the language of sexuality, the self and cultural community, effected by feminists in the 1970s and the gay community in the 1980s.[9]

And yet it is not just a matter of rethinking a certain condition, but of determining what such a rethinking might involve. How will it occur? Insisting on the "process of alterity" cannot mean the positing of otherness and of the celebration of differences as ends in themselves. Rather, what needs to be understood is that site of movement can be reformulated. Other potentials can be released. In order to approach these complex problems, three interrelated terms will provide the way through. The first is "symbols," the second "hospitality," and the third "cosmopolitanism."

Symbols

Music begins to play. A flag unfurls. An act of identification takes place. "We" identify with the flag. The flag identifies "us." Both moments have to occur in order that the flag's symbolic presence—its presence as a symbol—be maintained. The opening questions that have to be dealt with, and which are posed emphatically by this setting, are the following: Who are "we"? What is this "we"? In what way does the flag identify an "us"? These are the questions of citizenship—questions whose presence and thus ground of comprehensibility have to be rethought with the incursion of the refugee; the refugee as an insistent political reality; and the refugee in the broader sense of what is given within the interplay of location and movement. While the initial impetus within these questions concerns identity, the key to understanding them—and in trying to seek for what is disclosed by them—hinges on time. The time in question, however, harbors a certain complexity. There will be a twofold measure of time.

A symbol involves immediate recognition. Once seen, the symbol is recognized. At that moment, identification takes place and the attribution of identity occurs with equal force. Time in the first instance involves immediacy. Time, however, in the second instance will involve something else. The question of time concerns duration. For how long are we held and defined by the symbol? Does it announce an enduring truth about who "we" are? Or is the temporality merely strategic and therefore lasting only for as long as the symbol is held in view? While this may signal a concern that is merely abstract, the opposite is the case. "Our" being held opens up the important question of what, for example, an oath of allegiance would be like. And therefore, why the taking of such an oath may need to be made, and remade, on a continual basis. Identity is a temporal presence. Symbols are conventionally understood to have provided an enduring motif that invited and secured a group's identity—more often than not a national identity—for a sustained period of time. As a start, therefore, what has to be recognized is that any conception of identity—where that identity is secured and held in place—via the symbol is as much a question of the symbol as it is of time. A critique of the symbol, and thus a critical engagement with the conception of identity it envisages, and that is envisaged for it, also has to be a critical engagement with the temporality proper to such a conception of the symbol. Accepting this as a point of departure means that the opening description can be re-posed. What now exists is a type of frame.

As the music begins, the strains are recognized. The act of recognition ties music to memory. In a sense, it is collective memory. Everyone knows the tune; everyone can mouth the words. What is occurring here has to be given a description that is more than a simple redescription of the same thing. In order to move from a simple description, what has to be noted is the presence of a complex movement. Though this occurs as much with the music as it does with the flag as a symbol, it needs to be addressed from the beginning.

A flag unfurls. A sense of belonging is established or reaffirmed. The belonging in question, however, is marked by an ineliminable ambivalence. While these issues are well known, their reality warrants reflection. The flag could be a national flag, or it could be the one containing a football team's colors and logo. It could be a flag identifying with a certain political cause. The flag could be a national flag, though if it is carried in the context of a political demonstration in a country different from the one identified by the flag, this brings a more complex politics into play than those allowed by a simple nationalism. On the top of most town halls in Australia not only will there be a national flag, there will be the Aboriginal flag (accepting that there may be problems with any claim that this is the flag for all indigenous peoples), perhaps even a state flag or the flag of the local area. What is signaled by the copresence of all these flags? How is their symbolic dimension to be understood?

Part of the answer to these questions is that they symbolize two different—and at times related—states of affairs. The first is that there is a range of different subject positions that allows individuals to be individuated. Of equal importance, however, is the fact that these different positions can on a number of levels be in conflict. If the Australian flag is taken as commensurate with government policy, then the failure of certain key elements of reconciliation means that whatever is designated or symbolized by the Aboriginal flag stands in a state of sharp contradistinction to the Australian flag. With what do "I" identify? This is not an autobiographical question; though at a certain moment it may become one. The question in fact is: Who am I?

On one level the question has an obvious answer. The answer is given, for example, by the name that accompanies a signature; or by the name that takes responsibility for an action. The name begins to address the question: Who am I? Once, however, there is the need to defend that name, once that name is subject to identifications and reidentifications with which its bearer cannot identify, let alone control (examples here include being the subject of slander, libel, gossip, rumor), then the question of the nature of the "I" is robbed of its metaphysical as well as its temporal priority. If the "I" is a place marker for sites of activity, then what is meant by identification becomes a more insistent problem.

All of these questions raise the issue of how a proliferation of symbols can capture the complex reality of an identity that is always subject to process. Can there be symbols that hold "us" together? Or is it the case that they become merely pragmatic and thus simply strategic, and therefore do not have any actual symbolic value?[10] Once it is possible to see that what is at stake here is any sense of a conception of identity as immutable and that what has to be incorporated is a conception of identity—the identity marked and noted by the array of pronouns ("we," "us," "our," etc.) through which it is presented—then this inscribes the primacy of movement. The consequence of this inscription is that the refugee is no longer the exception but exposes the reality of "our" situation. Accepting this as a point of departure means, in this context, that inclusion or exclusion cannot occur if their justification is a homogeneous situation identified in the putatively unitary nature of the flag. This is why the position marked by the "refugee" becomes, in the end, though it was always there in the beginning, a location of the interplay between movement and settlement. However, precisely because questions of power are at work in borders and frontiers—sites held in place as much by legal frameworks as by imaginary constructs—movement through them, no matter where they are located, reinforces the necessity to define identity in terms of movement, though equally the ineliminability of power means that the negotiations allowing for admittance will be more than simple dialogue.

Hospitality

The question of hospitality arises because the refugee's pragmatic relation to borders is that the usual modes of entry and exit are either denied or, more aptly, have become redundant. The regulation of entry and exit once controlled by law—in the sense of national law as a series of statutes rather than the conventions of international law—means that movement is marked by passivity and regulation. From within this frame of reference, the possibility of a conception of hospitality as that which guides entry is redundant from the start. Surveillance and passivity would determine the response. The refugee, on the other hand, arrives without the constraints of acceptance as determined by statutes that would already be operable. Therefore, it is the refugee's insistent presence that causes the question of hospitality to be posed. In other words, it is only the presence of laws that already structure reception that allow for the question—how is it possible to be hospitable where that welcoming would not be governed by a preexisting legal framework?—to be posed. Once that presence is given priority, then what cannot be avoided is the issue of hospitality.

What does it mean to welcome? What is hospitality? There are at least two ways in which these questions can be answered. The first involves a mode of welcoming in which what is important is a sense of control. The welcoming is conditioned from the start. The conditions are established as much by regulations imposed by national laws—and in their imposition they simply reiterate the conceptions of sovereignty and place that are challenged by the refugee's insistent presence—as they are by conceptions of political diversity and practice. In contradistinction to this conception of welcoming, there is another one in which hospitality is not conditioned. And yet that does not mean that any sense of conditioning will be absent. Rather, what conditioning there is pertains, initially, to the one who enters. If an invitation is made and a door is opened, what is important is that if this sense of the unconditioned is to have any force, then while conditions are not set in order that entry be effected, once entry has been made two things follow. In a sense they are the same, though their entailments are different. The first is that the one who enters then has to live with the consequences both of having entered and of having been allowed to enter. Second, the one who allows entry—and here the "one" in question extends from individuals to states—without setting out determined conditions in advance, has to live with the consequences of having allowed entry to occur. What is defined is a conception of openness. Openness is linked to allowing—where allowing is defined as letting something happen. To allow is to create a space in which actions are not determined in advance and the consequences of such actions are not already evaluated in advance.

The important aspect of both positions is that from the very start they preclude the possibility that any action follows from entry, or that simply by virtue of having entered it would then follow that all subsequent actions are legitimate. Moreover, though this needs to be argued, the absence of formal criteria restricting entry does not entail that there cannot be a ground of judgment in terms of which the actions of the insistent refugee can and will be judged. The refugee, as the one who allows for unconditioned hospitality, opens up the possibility of judging the actions that follow from the conditioning of the unconditioned. This conditioning of the unconditioned is still prior to any regulation by law. Allowing is a space of responsibility and thus of action that is neither structured nor determined by legislation. (Though, and this would be the challenge, there is the possibility that it could inform legislation.)

Here, of course, is the link to entering—crossing a border, passing through a gate—within what could be described as an affirmative stance in relation to hospitality. The history of invasion and movement across borders has been marked by the attempt to take property or to redescribe property in terms of the possibility of possession. In the act of possession, the twofold structure of hospitality is necessarily effaced. While movement remains central, the settlement of the occupier is different than settlements structured merely by the constraints of sociality. The former is not defined by a generalized allowing, while the latter is. Entering and allowing to enter, here both under the guise of hospitality, have to be understood in terms of the link between hospitality and allowing. While it may seem overly complicated what this means is a certain commitment to a sense of openness. Both the terms "allowing" and "openness" do not entail, despite their appearance, the possibility that anything can occur. This is the position that has to be developed.

When formulating what he refers to as "absolute hospitality," Derrida writes the following:

> Absolute hospitality demands that I open my home [*mon chez-moi*] and that I give, not only to the foreigner [*l'étranger*] (provided with a family name and social status of a foreigner, etc.) but to the absolute unknown and anonymous other, and that I give a place to him/her [*je lui donne lieu*], that I let him/her come, and that I let him/her arrive, and to have a place in the place that I offer, without demanding either reciprocity (the entry into a pact) nor even their name. The law [*loi*] of absolute hospitality commands a break [*commande de romper*] with the hospitality of statute [*droit*] and therefore with law or justice understood as statute [*droit*].[11]

The importance of this formulation is considerable. While Derrida is right to insist on a distinction, if not a break, between law understood as statute *(droit)* and law understood as a type of obligation *(loi),* the opposition is too strict to capture what is at play in the idea of an entry effected under the heading of hospitality. Elsewhere, in pursuing the same topic, Derrida argues for a "pure welcoming," which would consist of "not only not knowing or to make as if one does not know, but to avoid any question concerning the identity of the other, of her/his desire, his/her rules, his/her language, his/her capacity for work."[12] While this would define a sense of purity and while it may structure one side of a relation of hospitality, the other side cannot be avoided. If movement is given centrality, then an entry that is a raid has to be distinguished from one in which refuge is sought. Even in the latter, the refugee can work with a sense of the unconditioned; namely, the entry cannot be disassociated from the consequences of having entered. If those consequences are detrimental to the refugee, then those allowing for entry work without the conditions set by hospitality; these conditions are allowing and openness. Reciprocally, however, the refugee has to be open to the possibility of hospitality. Once openness and allowing are given centrality, then on a philosophical level this means that the unconditioned is not in opposition to the conditioned; it is rather that there is a distinction between a conditioning defined by openness and allowing—and this may be able to take the name the "unconditioned"—and a conception of the conditioned where conditioning is the determining effect of legislation, be it progressive or not. On a political and hence pragmatic level—the level inevitably bound up with the enacting of policy—once openness and allowing are able to condition by defining hospitality, then the question would be: What is the correlate on the level of legislation to that state of affairs?

What can be concluded from this is that centrality cannot be given to the opposition between unconditioned and conditioned hospitality. Of far greater significance are the complex and differing ways in which conditioning takes place. Conditioning is not all of the same value. Nor are all instances of the same status or with the same result. The continuity of incursion and thus the continual necessity to think through the complexity of hospitality will have, or rather could have, the same structuring force in relation to the refugee's insistent presence as it would to a mode of relating to the other that is already present. Once the border condition is defined beyond the hold of the literal border, then the condition that determines the relation to the refugee will play a major role in the structure of relationality at every moment within human sociality. To the extent, therefore, that the frontier ceases to be just the literal border, then the truth dramatized by the refugee becomes the truth of human relations. The transformation of place beyond its reduction to the literal allows place to have histories, to have an ecology, and yet be traversed by structuring forces and lines of intensity that are not reducible to the simply literal. They are lines that in delimiting conflicting histories call, on the one hand, for the continuity of a form of negotiation with that conflict and thus with these histories and their related differing senses of historical time.[13] On the other hand, the refusal of the literal opens place beyond an oscillation between the international and the national. Precisely because it is a place to be determined, this interruption of determining oppositions—an interruption that refuses the form of utopia, since a connection to the literal place is fundamental—allows for place to be linked to potential. Potential is a release. The release in question is the cosmopolitan.[14]

Cosmopolitanism

While writing about the question of place, and the specific question of how, within the Australian context, place is to be understood, Stephen Muecke makes the following claim. (Within this claim the contrast with "North America" is intended as part of the question of what is involved in the delimitation of particularity.)

> To be an Australian I would assert means to imagine an anterior more than, for instance, the North American frontier. In a temporal sense this anteriority in indigenous ancient history, and the most recent history of conquest, are still abiding and unresolved themes of identity.[15]

On one level this argument evokes place as a literal place; the place that can be identified and mapped in terms of literal physical and political geography. However, it is the need to view it as part of an imagining that strips such a formulation of any easy evocation of the literal. In addition, once there is a move away from the possibility of a complete reduction to the literal, then this will allow for the setup in which particulars have a general dimension that is only ever actualized in terms of particularity. What this means is that place—perhaps what can be designated as "placedness"—brings with it the insistent presence of a real place along with an actual, though to be determined, history. This "placedness" is always more than or in excess of any simple connection to that place. The two, once taken together, work to redefine what is meant by both the literal and the imagined. The force of Muecke's argument is, in part, that it is an account of a setting that individuates and yet is, of course, a setting that is not settled.

What is intended by the expression "not settled" is twofold, and the folds are connected. In the first instance, it attests to the complexity of place once it is held apart from any quick reduction to literal place. The second is that what flows from refusing this reduction is the possibility of conflicting and disputed (and disputable) histories or narratives that are engendered by it.

Part of what also follows from this aspect of Muecke's position is that the resolution of identity will not be a simple matter. The question that emerges from this description is: What does it mean to negotiate this conception of anteriority? Not only will answers to this question differ in relation to how the negotiation is taking place, but it will also be the case that those differences will have an important relation to the position from which they are being established. What this means is that placedness involves conflicting relations and the history of conflicts. Neither one nor the other can be resolved by either the elimination of the conflict or the spurious strategy that would be linked to its having been overcome by its being forgotten. At work in the formulation provided by Muecke to the question of the placedness of Australian identity is a way, therefore, of understanding how the reworking of place is to be understood and therefore how it is to be undertaken.

What makes this question more than a local concern is that the conflicts in question do not just concern a straightforward relationship between an indigenous people and settlers. That relationship has a fundamental position. However, once movement and thus the presence of the refugee are brought into consideration, it becomes possible to redefine the local in terms of a diasporic condition. However, now the diaspora is not outside. It is rather that it involves what can be described as the "diasporization of the nation state."[16] While there are many ways of formulating this condition, one of the most important can be found in the way James Clifford defines a diaspora as bound up with what he refers to as "discrepant cosmopolitanism." He argues that

> the term "diaspora" is a signifier not simply of transnationality and movement but of political struggles to define the local, as distinctive community, in historical contexts of displacement.[17]

The significance of this formulation lies in the way in which place is positioned. The local becomes the unsettled and therefore place acquires a different quality. The possibility of settlement necessitates negotiation. While this is a concern that has direct political and ethical implications, it is one that intrudes into the fabric of the social. Precisely because of this intrusion, there are architectural and urban implications to this reworking of the local. In fact, to argue that there were not would be to conceive of both domains—architecture and urbanism—as untouched by questions of locality. While it may be the case that questions of this nature can be ignored, such acts, whether intentional or not, are predicated upon retaining place as the literal site. Allowing for a complex sense of locality is to allow both for this reduction to be held in abeyance and to reposition locality in terms of the cosmopolitan.

The contention throughout this chapter has been that the refugee reveals the truth about the situation in which modernity finds itself. While this is a uniquely modern concern, it is not as though the potential was not an already implicit possibility in the fabric of human relations—in the most abstract sense of self/other relations—or in the relation between movement and stasis. Rather than emerge as the exception, the refugee is part of an already present set of relations. However, that presence is as a potential. Rather than being the exception, the refugee is part of an already constituted set of relations. However, the potential is, by definition, not realized. Consequently, within this context its realization—which can have an effect as much on philosophical thought as it can on policy-making—will have a transformative effect. Once the reduction of place to its literal presence is no longer a possibility, once the "struggle to define" the local becomes that which guides policy—and again it has to be noted that it is the insistent presence of the refugee that will allow thinking to have a determining effect—and finally, once there has to be enactment of a locality where the local is not the national, then, among other things, this will have the result of causing a redefinition of site. To this redefinition there have to be added the complex temporal determinations in which indigenous temporality on its own will not suffice nor, for that matter, will the temporality of the international—the eternal now of the internationalization of capital. The interplay of place and time, once taken together, serve to interrupt the oscillation between the national and the international, since their inherent complexity opens up a space. Once this occurs, then any attempt to work with place has to take up these complex determinations. As such, urbanism cannot avoid this rethinking of place, precisely because ignoring its presence becomes either a refusal of the complexity of the present or a forgetting. The latter has at least two forms. In the first instance, it can occur in the service of a unitary conception of the national. In the second, it would be the result of a nostalgic conception of the indigenous. Allowing for the impossibility of these situations is a repositioning of site, as has been intimated, in the name of the cosmopolitan.

Architecture after the arrival of the refugee is not the architecture of accommodation or resettlement. (Though such possibilities, once reworked, can hardly be excluded.) "After" in this context means "in the style of." And yet, consistent with breaking with the hold of the literal, and therefore with a rethinking of place beyond those constraints, style cannot be reduced either to appearance or to the enactment of a single determination. Rather, it is an architectural practice driven by having to allow for a complexity of program, since this is what the question of community will now demand. Equally, it will need the invention of forms and of facades (the facade being the "appearance" of the building) that complicate the possibility of a single investment. Effect will become an architectural question once the present is allowed the intricacy that, in fact, marks it out.

"Intricacy," "complexity," and "complication" are some of the terms that are necessary in order to trace the impact of the refugee. What has to be countered are the claims that, first, the arrival of the refugee cannot occur within a structure of hospitality, since the law does not allow it; and second, the arrival of the refugee has to be met by arguments for the homogeneity of national culture. Responding to both these assertions is to affirm the position—a position exposed by the refugee—that, in the first instance, hospitality cannot be restricted by statute, but that statutes can be redefined by allowing the structure of hospitality to have a determining effect on their constitution. And second, that national cultures are, and always have been, potentially or actually diverse. Diversity has not just to do with ethnicity; it also concerns the differing temporalities of arrival and settlement. All are present at once. Hence, the singular time in which they are present becomes the complex times of the present. In relation to this setup, the refugee is not the exception but a further element. Allowing for that extra addition involves redefining addition in terms of movement. The refugee becomes another movement; one of a potentially infinite number of movements through or within a site. The norm is to view the refugee as the exception. Allowing for the centrality of movement means that what is exceptional is that the "norm" is treated as aberrant. While this has to be the case, since the defense of borders and frontiers in the name of a unified national culture depends upon it, arguing for the incorporation of the refugee—and thus for hospitality—cannot have recourse to a position that would maintain the refugee as the aberration such that charity toward the aberrant became all that was possible. (Among other things, this fails to question the idea of a unified national culture.) The contrary is the case. The refugee, in its emphatic form, is the truth of our condition. Accepting the refugee, accepting as a mode of hospitality with all that it entails, involves a difficult series of negotiations, ones with an impact that registers as much in the realm of philosophical thinking as it will in the domain of architecture and urbanism—the terms designating a formal response to place—that would stem from accepting and affirming "our" condition. That condition, while always located within the confines of the nation, transforms place and thus transforms the nation. Transformation does not mean a redefinition in terms of the international. Once the transformation is strategic—and therefore bound by repositioned local concerns—then "our" condition becomes the cosmopolitan.[18] As such it becomes a local concern.

7. The Matter of a Materialist Philosophy of Art: Bataille's Manet

Édouard Manet, as Clement Greenberg argues, is "exceptional in his inconsistency."[1] Equally, he is, in different guises, linked to the moment where painting becomes its own concern. Modernist painting, it can be argued, is a project that begins to be worked out with Manet. Greenberg is right about the "inconsistency." The question, however, is how that state of affairs is to be understood. Identifying the particularity of the project of modernism with Manet, while a viable option, also gives rise to the problem of how such a conception of periodization is to be understood. The question is straightforward: What is at stake in these claims about Manet? This is, of course, part of the opening toward Bataille's Manet.[2] For Georges Bataille, the name Manet has "a meaning [*un sens*] apart" (*OC* 9:115).[3] He has, moreover, "broken with those who preceded him" (*OC* 9:115). How, then, is this "apartness" and thus this breach—this cut—to be understood?

And yet it should not be thought that painting's concern with its own presence as painting dates just from Manet. There is a continuity of such a concern. However, it is a continuity that is discontinuous in terms of its own self-presentation. Nicolas Poussin's famous self-portrait of 1650, the one executed in order to be sent to his patron Paul Fréart de Chatelou, has been the subject of a number of detailed analyses, most of which have been concerned with the painting as a scene of representation. This particular work is of special interest because it follows another self-portrait done the preceding year for John Pointel, thereby implicating the question of the self-portrait in the broader question of painting's own self-presentation.[4] Rather than pursue the relation between the two self-portraits, in this instance greater significance will be attached to the way the 1650 self-portrait announces its presence as painting. The painting's organization—its use of light and shadow, for example—has been linked to a series of rhetorical tropes.[5] However, identifying the painting's importance as a site of self-reflection is best situated in relation to an element of its initial reception. Here Giovanni Pietro Bellori's 1664 account in his *Lives of the Modern Painters, Sculptors, and Architects* is central.[6]

Poussin's self-portrait of 1650 is distinguished not just by its incorporation of scenes of representation—canvases, frames, walls, books—but by its use of these scenes to site its own self-presentation of painting. This occurs within one of the framed works located in the actual painting. Bellori identifies the way it takes place without equivocation. Within the painting, on one of the canvases can be seen part of a woman's head. Of the woman in that picture—the picture within the picture—Bellori writes, "questa è la Pittura."[7] She is the allegorical figure of painting. Already an allegorical figure, she neither symbolizes nor signifies painting. She is it. In other words, here in a painting is painting referring as much to its own source as to its own possibility. This occurs by the incorporation of the figure of Pittura. Leaving aside the iconic reference to friendship—present in part in terms of the hands reaching out to Pittura's face—what has to be addressed is the way painting figures within the painting.[8] The distinction is best captured graphically by distinguishing between painting and Painting. What this distinction is intended to capture is that painting is present neither as act nor as technique but only, and exclusively, as an allegorical form. The form is present within and as a work of art, which is, moreover, juxtaposed with other sites of representation.

The consequence of this picturing of painting is clear. The way that allegorical presence is present—namely, due to the act of painting and the employing of techniques not necessarily articulated within the tropes of rhetoric—has to subordinate any interest in techniques to the painting of the figure. The figure therefore is the place of painting. And it takes the place of painting, were painting it to be considered as an act, rather than that which has already been enacted. (Hence Painting rather than painting.) The form of the figure will always have been present. Its form is already determined. Once this already-present determination has priority, then there is the twofold move in which what is included is a form whose determinations have been given in advance—this would be the condition of her recognizability—and what is excluded is the possibility of identifying painting as a concern with *forming*. The move that would give centrality to forming—a move for Bataille that is marked by the name Manet—involves a repositioning of painting consequent on emphasis now having been given to the act of production. Painting emerges as the specific activity of form creation. In Poussin's 1650 self-portrait, painting is always linked to its capacity to represent itself. However, as has been suggested, what is represented defers the question of painting as an act, by representing itself as a figure. Figure as the displacing of the act—i.e., the displacing of painting as *forming*—defines the pictorial space as representational precisely because it is the clarity of the figure's representation, and thus its representational presence, that secures painting's self-presence. These tightly knit interrelations define what can be described as the hold of representation. Central to this is the definition of technique in terms of representation. (Hence the formulation starting with Quintilian, though also present in Bellori, of artistic techniques in terms of rhetoric.) What Bataille notes in Manet are, first, the break of that hold and, second, the corresponding redefinition of technique. In other words, it is not simply that painting will no longer be defined in terms of the success or failure of representation; it is also the case that technique will no longer be thought in terms of that which successfully (or not) enacts painting's representational power. The place of representation has an inherent complexity. Representation is not reducible to thinking of painting in terms of the re-presentation inside the picture frame of that which was, by definition, exterior to painting. Overcoming the hold of representation stems neither from a discursive critique of representation, nor the elimination of presentation within the field of painting; rather, it is the productive distancing of a conception of pictorial space defined in representational terms. What matters is what appears. What appears is, of course, the appearance of matter.

Overcoming the defining hold of representation—in this extended sense of the terms—means that part of the answer to the question of precedent, and therefore of Manet having "un sens à part," is given in the relationship between matter and appearance. (A relationship that already incorporates an operational practice; i.e., technique.) More generally, what is at work in this relationship has to be linked to a specific conjecture about artwork in the modern period. A conjecture which marks, in part, the cessation of the hold of representation as necessitating either the deferral of technique or its having to be defined in terms of representation. A more exact way of stating this concern would be to designate it, as has been intimated, as the question of appearance in the modern period. Indeed, the argument has to be that what marks out the production of the plastic arts, as well as architecture, within modernity is the separation of style and appearance. (In other words, there is no one appearance proper to the modern. Even for Greenberg, flatness, for example, could not be directly equated with abstraction.)[9] One of the most emphatic originating moments of this separation is Hegel's identification of the limits of art. The limit is there in art's definition; a definition that is as much involved with the place of art as delimited by its being the sensuous presentation of the non-sensuous, as it is in the separation of the art object from its immediate understanding and therefore, as will be noted, from immediacy in general. In regard to the latter, the space opened by the separation becomes the site where both philosophy and criticism can be situated. It is, of course, neither simple nor mere sensuousness, as though all that was present was a mere empirical object.

In the *Lectures on Aesthetics,* Hegel clarifies this point in the following terms:

> In the sensuous aspect of a work of art [*im Sinnlichen des Kunstwerks*] the mind seeks neither the concrete framework of matter, which desire demands, nor the universal and merely ideal thought. What it requires is sensuous presence, which, while not ceasing to be sensuous, is to be liberated from the apparatus of its merely material nature. And thus the sensuous in works of art is elevated to a mere appearance [*blossen Schein*] in comparison with the immediate existence of things in nature, and the work of art occupies the mean between what is immediately sensuous and ideal thought [*und das Kunstwerks steht in der Mitte zwischen der unmittelbaren Sinnlichkeit und die ideellen Gedanken*].[10]

This middle or "mean" position stands out from what will become the problem of the object, i.e., the reduction of the art object to the immediately sensuous, where that reduction has to be allowed for and then overcome such that an object becomes art.[11] At the same time, the artwork is counterposed to the ideal. Its inadequacy is given by the counter-position. Indeed, the centrality of philosophy and religion emerges because of their respective, and adequate, relation to the ideal. This separation means that art is given at least two options. In the first instance, to be viewed as always inadequate in relation to some form of ideality, or, second, to be reinvested with a force that allows it to take over a renewed sense of ideality. These two different positions yield, in the first place, Hegel and, in the second, Baudelaire. They are the thinkers of art's work within the opening separation of style and appearance. However, in regard to the formulation of this position, it is a determination of that separation that takes place in relation to an ideal. The significance of Bataille, and therefore of what allows him to emerge as a sustained thinker of modernity, is not simply the refusal of that determination—i.e., one that is inextricably defined in relation to the ideal's effective presence—but the way that refusal is enacted. As such, there is an important confluence between that enacting and the art of Manet. Manet could only be the painter of modern life if that life is invested with categories of the modern, categories as much ontological as temporal. The modern—the thinking of modernity—is not just a matter of history. What matters is the direction of thinking.

Baudelaire's argument, one that appears in different forms within his poetry and his prose, that Beauty—understood as art's ideal—is always relative to both a people and a moment of historical time, maintains the ideal. Its retention gives force to his claim that "Romanticism" is "the most recent and most modern expression of beauty."[12] The link to what has been argued above is twofold. In the first instance, criticism occurs in the separation of the ideal and the contemporary. In fact, it would be the task of criticism to establish that precise connection. And yet, on the other hand, the difference from Hegel is that the contemporary is the contemporary form of the ideal. Even though not devalued as such, the contemporary in Baudelaire is still defined in relation to the ideal. With Bataille, there is more than a simple reversal. In other words, it is more than a redefinition of art. The modern will involve not just an interruption but a fundamental shift in the categories within which the event of art is to be understood.[13] On this point there is an important affinity between Walter Benjamin and Bataille.[14] For the former, it is not simply that technical reproducibility causes an interruption. It does. Of equal importance, however, is that the interruption introduces, as part of the event of art, a change in the concepts and categories that are demanded by the shift in the nature of production, but equally one that can then come to be deployed in relation to the interpretation of art in the periods prior to technical reproducibility. In fact, for Benjamin it is that opening up—one in which the *Nachleben* of works is activated—that then allows them to be historical. Decontextualization becomes the sine qua non for the production of historical objects.[15] This is not Bataille's argument. Nonetheless, the affinity lies, in this instance, in the recognition that historical periodization—a formulation that covers the different ways in which the advent of modernity is thought as an interruption—has as much to do with historical time as it has to do with categories of thought.

Bataille and Materialism

Manet's complex and critical relation to the history of painting is well documented.[16] That relation does not simply dramatize the question of precedent; it turns that question, as has been suggested, into one in which what is at stake is the appearance of modern art. While modern iconography continues to read in his work the symbols of the modern—captured, in the case of *Olympia* (1863), by an insistence on the interplay of prostitution and adultery—what is of interest here is what happens when painting is approached neither in terms of its having been determined by precedent, nor, perhaps more importantly, when it is construed as a field of ideas.[17] How is the matter of painting to be understood? After all, matter emerges not in the move from the idea to the empirical, but in the rearticulating of that relation such that the "ideational" content is an effect of matter. This is the opening in which Bataille's work on Manet, and with painting in general, can be situated.

Bataille highlights Manet's "indifference." And yet this indifference is neither disdain nor distance in any straightforward sense. It is a situated indifference, one which has, to use Bataille's terminology, "une force" whose presence accounts as much for the pleasure that can come from his activity as a painter, as it stages the works' destruction of the mythology that forms the basis of paintings by Raphael and Titian.[18] The destruction of ideas does not occur merely by the presentation of a different set of ideas; it also has to figure as technique. As will be argued, this destruction and the impossibility—or at the very least a complex possibility—of a relation between beauty and modernity are essentially interconnected. The works of Raphael and Titian, in being taken up by Manet, are neither the subject of borrowing nor of direct citation.[19] They were present within an indifference whose results were destructive. This is the contention concerning Bataille's encounter with Manet. Moreover, it is a contention that has to be situated within the framework provided by Bataille's conception of materialism.[20]

In a number of early articles originally published in *Documents*—the magazine which appeared in 1929 and of which he was the "secretary general"—Bataille took up the question of materialism and of what he identified as "base materialism." The significance of his work on Manet, which equally resides in his articles on Goya and Van Gogh, lies to a considerable degree in the relation between the work on materialism and the engagement with works of art.[21] Despite Bataille's own reservations, if not hostility, toward "ontology," his project is to develop a formulation of materialism that is not just a simple negation of idealism. Part of the force of materialism is its capacity to sustain a position that eschews representation, if that is understood as governed by an external determination, and thus falls beyond the purview of an aesthetics (let alone a philosophy of art) delimited by the failures of representation.[22] With Bataille, the fate of art has no inherent connection to either loss or lamentation. Furthermore, there is no given connection to a conception of impossibility that is defined, by the retained effective presence of an ideal, as a failure.

The two texts in which Bataille's understanding of materialism is presented are "Materialism" and "Materialism and Gnosticism." These texts were published in 1929 and 1930, respectively. The first of these texts—one of those comprising the famous *Critical Dictionary*—is concerned with outlining the need to move away from an idealist conception of matter. Such a conception is preoccupied with what Bataille describes as "an *ideal* form of matter" (*OC* 1:178). The move away, and thus the possibility of an approach other than a simple refusal or negation of an idealization of matter, is found in his insistence on an understanding of materialism as a word "designating the direct interpretation of brute phenomena excluding idealism" (*OC* 1:176). While there is a temptation to see in this reference to "brute phenomena" the promulgation of an empiricism and not a materialism, it is Bataille's reference to Freud that immediately modifies, if not precludes, such a reduction. From Freud what has to be "borrowed," argues Bataille, is "a representation of matter." Two things mark Freudian materialism. The first is the presence of a dynamic process and thus an active or productive sense of force. The second is that the account of "presentation" in, for example, the 1915 paper "The Unconscious" allows for the formulation of mental activity in terms of an economic model and thus of a conception of presentation occurring as the result of the operation of forces.[23] Not the play of indeterminate forces, but the directed and complex interplay between the conscious and the unconscious in which already present mental images are mediated by psychical activity. The unconscious is not the site of non-sense, but of a different conception of sense. While Bataille does not draw upon this aspect of psychoanalysis, it is one that is, nonetheless, far from antithetical to his overall project. This latter positioning will have a fundamental effect on how the process of presentation within psychoanalysis is understood. Rather than its being linked to a theory of expression, "presentation" becomes the effect of a system and therefore has to be defined in relation to force working through the presence of already existing images.

The second of these early papers—"Materialism and Gnosticism"—offers the more sustained formulation of materialism. Bataille's fascination with Gnosticism is not linked to a concern either with the religious or the mystical. The approach is far more exact, perhaps clinical. While Bataille argues that Gnosticism provided that which conformed the least to the "established intellectual order" (*OC* 1:223) and was moreover a body of thought that never assumed the "role of the state religion" (*OC* 1:224), more is involved than simply an oppositional presence. The value of Gnosticism was that it could not be defined in terms of the oppositional logics within which bodies of thought as well as political orders were articulated. Its use lay in its uselessness. Strategically, what this meant was that Gnosticism was associated with a conception of matter that was not defined in opposition to the ideal, nor was it present as a degraded form of the ideal. The project of overcoming a positioning of matter in terms resulting from either a form of degradation or a process of idealization opens a space in which a reconceptualization of matter and of "le mal" can occur. (A spacing occasioning a politics of place.) This "mal" is not the counter-position to the good. It is a conception of "mal" that, to use a formulation he deploys elsewhere, "demands an ultramoral position" (*exige une 'hypermorale'; OC* 9:171). The question of indecency will emerge as integral to any treatment of Manet's *Luncheon on the Grass (Déjeuner sur l'herbe)* as well as *Olympia.* What is important, following this lead, is that it is not the indecent in opposition to the decent.[24] In fact, the base can be situated outside the hold of the movement between the material and the ideal. This movement is not a simple opposition—material/ideal—since while one is defined in opposition to the other, it is also the case that one can be defined in terms of the other. The attraction of psychoanalysis was that it started with matter as articulated within a dynamic process.

Bataille begins to formulate this other position—a position already involving another politics of place—in the following terms.

> Base matter [*la matiere basse*] is outside and foreign to the aspirations of human ideals and refuses to let itself be reduced to the large ontological machines resulting from those aspirations. (*OC* 1:225)

Ontology, it has to be added, can return once it holds itself apart from a definition given by the uncritical acceptance of the relation between universal and particular. This "exteriority" of which Bataille writes, while worthy of a detailed description in its own right, has important implications for art. In regard to Gnosticism, this resulted in what he describes as a "figuration of forms" that "contradicted" academic demands. Contradiction was not a simple negation. Rather, the image of base matter appeared in ways that were incongruous with, or were without regard for, the propitious demands of idealism. This is the link. Bataille concludes his reflections on Gnosticism with the argument that "today,"

> in the same way [*sens*], plastic representations [*figurations*] are the expression of an intransigent materialism with recourse to everything compromising the established powers in matters of form, ridiculing traditional entities, naively rivaling stupefying scarecrows. (*OC* 1:225)

While Bataille does not intend it in this way, it is a formulation of this type that allows for a differentiation, within French art, between David and Manet to be drawn. David's relation to classicism, and in part this is the strength of David, is that he has to use classical form in a way that leaves it untroubled, in order to allow Rome to be that through which the image of secular republicanism can be presented and, from within the activity of painting, imagined.[25]

In other words, within Gnosticism Bataille is able to identify a mode of thinking whose contemporary purchase does not lie, in any direct sense, in the history of religion but in the way that it promulgates a conception of materialism that is recalled by certain activities within the domain of artistic production.

Bataille's Manet

It is this evocation of the "plastic arts" that allows for a direct entry into Bataille's text on Manet to be made. Manet's work will be bound up with an "intransigent materialism." Moreover, what will also be recalled is the presence of "brute phenomena," now linked to a form of realism positioned beyond empiricism as a philosophical position, with the art of Courbet as its artistic expression, and the writings of Zola as its literary enactment.[26] For reasons of brevity, concentration will be given to Bataille's interpretation of *Olympia* (1863). For Bataille, this work brings into play that which is essential both to Manet and to painting in the modern period (*OC* 9:137). Perhaps the most emphatic moment in which the concerns of "base materialism" and *Olympia* coincide is in his use of Paul Valéry's judgment of the painting, in which he uses it almost contra Valéry.

Valéry, in his "The Triumph of Manet," writes that with *Olympia* there is the presence of the "impure" itself. She is the "absolute nude." Valéry goes on to suggest that "she provides a way of thinking about everything that is hidden and conserved of the primitive barbarousness of ritual animality in the costumes and the works of the prostitutes of the large cities" (*OC* 9:142). Valéry interprets the painting in terms of what it gives. Bataille understands this gift as the "text" of *Olympia.* In other words, it is the presence of painting as text and not as the work of painting. Bataille, as will be noted, wants to argue that what the painting "signifies" is the effacing of the text. The appearance of modern painting is not the appearance of content; rather, it is the "song of forms and colors" (*OC* 9:126). Technique has become central. Rather than a preoccupation with either meaning or acting by the absorption and transmission of tradition—in either positive or negative terms—there is an overcoming of "sens." (Overcoming is, of course, not the nihilism of pure visual destruction or the naïveté of disavowal; art as anti-art.) With *Olympia,* what is, is that she is there. And she is only there because of technique. (Technique, were it to be detailed, would have to show how the positioning within the painting defers as much to the history of photography as it does to the use of brushstrokes and the realization of bodily presence.) The question therefore has to concern that presence. Impersonality predominates such that she is not there as the embodiment of anything other than herself. The simplicity of her presence is what can only be described as the interplay of absence and silence. Contemporary criticism—criticism at the time of Manet—which saw in her the "ugliness" of a "gorilla," Bataille accounts for in terms of the painter's concern "to reduce what he saw to dumb simplicity, to the gaping simplicity of what he saw." This reduction, which is of course not a reduction that debases, has to be positioned as base in the precise sense of a baseness beyond the hold of morality. Here, therefore, is the commitment to realism. It is, however, not the realism of Zola, since that realism was linked to the situation and thus to site.[27] The realism of Manet, for Bataille, is that *Olympia* could be situated "nulle part." This has to be understood as the claim that *Olympia* demythologizes beyond the hold of myth. She is present as a woman and not as a Goddess. She embodies only herself. This accounts for why Bataille adds in regards to the work as a whole: "Everything in her slid to an indifference to beauty" (*OC* 9:147). The slide involves neither refusal nor disavowal. (Nor, moreover, is it the ugly.) In this context, "indifference" becomes the identification and the affirmation of the base. It can, however, only be affirmed once the oscillation between the ideal and its other no longer creates the setting delimiting the locale of criticism and thus defining the reach of interpretation.

This claim can be clarified further since what it entails is that *Olympia* does not stand for an ideal that would be counter to an idealization of woman as "Goddess"; equally, however, she is not merely the negative counterpart; this accounts for Bataille's insistence that she does not have a place. If there were a question posed by *Olympia,* it would concern the possibility of a place, not for her in the sense that she should be given a home, but the possibility of place housing the base and accommodating those defined by "le mal." Where are they to appear? This question, as has been noted, is the one that allows for a link between art and politics. The political question is raised, not resolved, by art. Pursuing the politics of art could usefully take up the opening provided by Bertall's caricature of *Olympia.* Not only does this indicate the limit of satire, but at the same time it indicates that caricature has to miss what is at stake in art. The reduction to a redescription or re-presentation of characters has to reduce the artwork to the site where characters—ones having an inevitable even though banalized form of stability—are then able to sustain caricaturization.[28] In other words, it has to rob art of its informal nature by determining it in advance as a field where people—subjects—are active. (This will be the case even though they may be presented as objects of derision or even hatred.) And yet, what satire does indicate is the inherent fragility of art, particularly an art form whose relation to public taste is complex.

In a general article on impressionism written at the same time as he was preparing the book on Manet, Bataille addressed the problem of this fragility. In the first passage cited below, fragility is present in terms of "sens." In the second passage, fragility is present as subtlety of subversion, and here it has to be the subtlety and subversion of *la folie.* Subversion does not need the excess of literal violence.

> It is only from Manet that to the conventional aspect of things, painting substitutes an unforeseen [*imprévu*] aspect which does not have sense [*sens*] other than the incongruity of things as they are [*comme elles sont*]. (*OC* 12:373–74)

> It represents a broken movement, an often unexpected [*inattendu*] eruption in domains where plays an element of disorder and subtle subversion. (*OC* 12:375)

Three elements mark the first passage. All are linked. The first is the "unforeseen" *(imprévu);* the second is "sens"; and the third is the commitment to realism announced in the words, "things as they are" *(des choses comme elles sont).* The "unforeseen" is a term that reappears elsewhere in Bataille's writings as the "l'impossible," where it is defined in opposition to "the object of inevitable choice" (*OC* 3:511). After Manet there is a need to respond to the "unforeseen." The argument is that the question of criticism emerges as a question precisely because of the inescapability—inescapable if attention is paid to the demand of art—of that need. Moreover, the "unforeseen" demarcates the deferral of already determined forms and thus defines criticism in relation to technique and therefore to the process of forming. (The latter always has to be understood as the finding of form.) Again, the assumption is that content cannot be divorced from the centrality of technique. Once "sens" is denied its already determined place—though it is a "sens" that resists nonsense—place will always be fundamental. In no longer being given either as at hand or there in a utopian projection, place actualizes an inherent politics of meaning. The "unforeseen" is inextricably bound up with the event of art. With that actualization, a further critical task emerges. For Bataille, these challenges can be pursued precisely because another direction has arisen. "Sens" is no longer given in advance. Here is the informality of realism.[29]

In the unpublished notes to his book "The Impossible," Bataille describes it in the following terms:

> No hidden meaning. [*Pas de sens cache.*]
> Above all no system. [*Surtout pas de systeme.*]
> Insist on the realist character of things. [*Insister sur la caractere réaliste.*] (*OC* 3:509)

What is at work in these notes is more than an opening toward other possibilities; they are defined beyond the mystical and its celebration of a community of one. Hence neither meaning nor directionality is "hidden." Equally, the incorporation of a particular in a "system" is also not an option. This does not mean that there is only the idiosyncratic. Rather, the overcoming of system is that which allows for the "unforeseen" to interrupt and thus occasion an opening. The "unforeseen" cannot be accounted for in terms of the "system" that generates it. (Hence, the nature of the realism.) The "hidden" and the "system" stand opposed to the real. "Insisting," while strategic, is also subtle. This is the link between the realism announced in these notes and the "subtle subversion" in the second passage noted above. This subtlety is reiterated elsewhere in terms of latency. When writing of Manet's painting *The Fifer,* and in arguing that the work needs to be explained in terms of "a resolute exposition of technical possibilities"

(*OC* 9:156), Bataille adds that this is achieved with an elegant fury. Never fury without the qualification of elegance.

What is reiterated here—and the repetition has to do with form and not content—is Bataille's conception of poetry as that which is positioned by a relation between the "known" *(connu)* and the "unknown" *(inconnu).*[30] If poetry were only the former—the "known"—then it would be either information or the return of that which was given to be repeated. In other words, it would be a repetition of a determined content and not of a formal possibility. If poetry were only the latter—the "unknown"—then it would have become mysticism or a hermeticism not allowing any form of community. While it is the power of the negative that retains its hold—poetry's refusal to be known and therefore to elicit a response, exerting its hold through the impossibility of its masterability—poetry also works through the flecks of the known.[31] Toward the end of the text on Manet, Bataille argues that "what counts in Manet's canvases is not the subject, what counts is the vibration of light" (*OC* 9:157). On its own this point would be banal: the banality he attributes to Malraux, who makes a claim of a similar nature. What matters is how the "light" is understood. For Bataille, it is the sacrifice of the subject that is maintained by it. Moving from the language of sacrifice, the material effect returns as the event of art.[32] *The Fifer,* as subject and as work, is the effect of matter's work.[33] Its presence can be separated from the continuity of its production. The event of art is bound up with its production.

Forming

Finally, therefore, what emerges from this consideration of elements of Bataille's encounter with Manet is the centrality of a series of originating terms, all of which indicate that the move to form has its point of departure in the informal. "Apartness," "the unforeseen," that which stands counterposed to "inevitability," all link form to a relation of distance. Openings emerge not as the consequence of the positing of destruction bound up with new beginnings, but through acts—the event of art—that are only ever confirmed by the work of criticism. What this means is that Bataille's Manet can be read as a response to the question: How, given the work's refusal of continuity and public taste's refusal to confirm its presence as art, is that presence to be established? Even accepting that establishing that presence has an inescapable contingency, it should not be thought that all that can be done is to establish conditions of relativity.

Bataille's conception of materialism allows for an account of a realist conception of art practice to be given. In such a context the real is situated beyond the oscillation between the ideal and either its physical enactment or its ruination. To the question—what is the real?—the answer will always be, what is true. In his article "Hegel, Death and Sacrifice," Bataille quotes from Hegel in order to establish the connection between the real and the true. "Spirit only attains its truth in finding itself within absolute sundering" (*L'ésprit n'obtient sa vérité qu'en se trouvant soi-meme dans le déchirement absolu; OC* 12:331).[34] In another article—"The Paradox of Eroticism"—this formulation of the truth of subjectivity is linked to literature and by extension to art. In a remarkable passage, Bataille formulates this position in the following terms:

> In the tearings [*les déchirements*] to which the miracles of our joys take us, literature is the only voice, already broken, that gives us to this glorious impossibility where we are not being torn, it is the voice we give to the desire to resolve nothing [*au désire de ne rien resoudre*], but visibly, happily of giving ourselves to a sundering [*déchirement*] without end. (*OC* 12:325)

Art becomes that through which the truth of subjectivity can be staged. More is involved, however, since what is at work is the structure of thought within which this conception of the subject and art is formulated. The "déchirement" is not an undoing; it is an original state. Yielding to it becomes the enacting of a founding truth whose occurrence is prior to unity or fixity. The fixed is always there in the moment after. The fixed is form. (Reinterpretation, which is the potential of artwork, indicates the extent to which the quality of informality also resides in fixity.) Locating fixity as given in the relationship between the informal and the formal entails that what is at work in Bataille's use of Hegel is not reducible to claims simply about subjectivity. What is involved is a claim about the truth of form; namely, that it is inscribed within a process and thus best explicated in terms of what has already been identified as *forming*. What is being worked out with Manet is forming. Terms such as "apartness" and the "unforeseen" provide a way of accounting for resolutions which, while definite, carry with them what Bataille identifies as the "desire to resolve nothing" and thus what was referred to earlier as the power of the negative. The opening marked by this desire is one where an aesthetics could emerge as the result of a demand being made. However, it would not be a demand for literal destruction, but for one which carried with it the truth of subjectivity and thus for the continually irresolute. What is recalled here is, of course, Bataille's description of *The Fifer.* The site of continual irresolution is formal. (A finality allowed by the impossibility of *forma finalis.*) The resolved, in this instance, has to do with the technical possibilities that make the subject the effect of the work; its having been framed. The erasing and reinscription of the subject matter—a reinscription premised upon the removal of one conception of what has priority and attributed centrality, hence the evocation of sacrifice—entails that content has been stripped of any symbolic value. As such, it becomes an image beyond the hold of the symbol. This beyond is not absolute. The already identified place of fragility endures. Content can always become the site of a reinvestment. In other words, the intrusion of a symbolic dimension—turning the fixed into a narrative imbued with a content becoming text—always endures as a possibility. And yet, it is not as though *The Fifer* does not involve, for example, a certain conception of youth. The point is that it is an effect of technique and not a positioning explicable in terms of icons or symbols. Moreover, precisely because it resists the evidential status of a photograph, it should not be construed as finally resolved. A resolved formal determination of content matter could never be the province of artwork. After Manet the extension of this position becomes irresistible.

What Manet's work dramatizes is artwork's ineliminable aleatoric quality; a quality that makes impossible the attribution to art of the forms of resolution and determination that are external to it. (This is art's autonomy, an autonomy also captured in art's irreducibility either to information or to its complete inscription within a logic of exemplarity.) With Manet, art works outside the constraint of final form and therefore is bound up with the continuity of the question of technique. This is a question that will eschew any reduction to the merely technical, insofar as it is technique that will inform the activity of criticism. The effect of this opening and the incorporation of technique is that it becomes possible to interpret artworks in terms of forming. Forming, however, has not merely emerged. It is not as though it is simply posited as a possibility. In emerging from art's history, it allows that history to be recast. This is not to essentialize artwork, but to indicate the extent to which its history is inseparable from the practice of criticism.

Notes

Introduction

1. The references to Walter Benjamin's texts in the original German are to the *Gesammelte Schriften,* ed. Rolf Tiedemann and Herman Schweppenhäuser (Frankfurt am Main: Suhrkamp, 1980–91), which is abbreviated as *GS* in the text and notes, and is referred to by volume and page number. References to English translations of Benjamin are chiefly to *Selected Writings,* ed. Michael W. Jennings (Cambridge, Mass.: Belknap, 1997–2003), which is abbreviated as *SW* in the text. There are also references to *The Arcades Project,* trans. Howard Eiland and Kevin McLaughlin (Cambridge, Mass.: Harvard University Press, 1999); if possible, these references will be to the Konvolut numbers rather than the page number; if a page reference is necessary, then it is abbreviated as *AP.* References to Benjamin's *The Origin of German Tragic Drama,* trans. John Osborne (London: Verso Books, 1998), are abbreviated as *OT.*

2. Hegel deploys the term the "cunning of reason" in section 36 of his *Philosophy of History.* There is no contention here that Benjamin is involved in a detailed negotiation with Hegel which would be limited simply to the use of terminology. Rather, Benjamin's position is constrained to refuse Hegel's conception of history as the unfolding of Spirit; moreover, Hegel views history as an unfolding that involves an inherent commitment to a conception of history as progress, even if it is an idealist conception of progress. In addition, the specific term "cunning of reason" identifies the fundamental accord at the basis of discord. Benjamin's use of "cunning" can be read as undoing both this conception of historical temporality (in sum, teleological progression) and the actual structure of conflict. There is an affinity here between Bataille's critical engagement with Hegel and Benjamin's. The point of departure taken by both—a departure that deprives Hegel of any real utility, even if that utility would come from an inversion of his position—is the structure of idealism. To invert Hegel, Bataille suggests in certain texts, is to do no more than generate an idealist conception of materialism. See Georges Bataille, "Le matérialisme et la gnose," in *Oeuvres complètes* (Paris: Gallimard, 1970–84), 1:178–80. All future references to Bataille in text will be to the *Oeuvres complètes,* which is abbreviated as *OC,* with the volume followed by the page number.

3. Karl Heinz Bohrer explores the relationship between interruption and utopianism in *Suddenness* (New York: Columbia University Press, 1998). The key moment is his interpretation of the famous "madeleine" passage in the first volume of Proust's *Remembrance of Things Past.* His argument is that

> this passage presents the first element of the utopian "moment": the pure self-presence of the "now" in the temporal setting of the instant. (199)

The other possibility, of course, is that if such a moment is productive, then rather than giving an account in terms of the positing of another place, a more adequate one would be given in terms of the interruption's potential. It is this latter possibility that implicitly informs my understanding of interruption in Benjamin, though it is explicitly present in the engagement with place in the second half of this book.

4. As will be argued in chapter 1, the "Absolute . . . is that which allows for the interruption, but equally it is what is evidenced by it." A further reference can be made here to Hegel. For Hegel, there cannot be a moment that is a pure singularity and thus that does not form part of a whole. See in this regard the argument against singularity in the *Shorter Logic* (Oxford: Oxford University Press, 1972). There are many passages that could be referred to substantiate this point. One of the most straightforward is in section 13. Here Hegel argues that despite the presence of diverse and divergent philosophical systems, this appearance is not the reality since

> the same architect has directed the work: and the architect is the one living mind [*ist der eine lebendige Geist*] whose nature it is to think, to bring to self-consciousness what it is, and, with its being set as object before it, to be at the same time raised above it, and so to reach a higher stage of its own being.

The singular, therefore, is only singular on the level of appearance. Counterposed to the singular as an end in itself is what Hegel means by "system." However, the undoing of this position does not mean the positing of pure singularity. (This would be to attempt to counter idealism with empiricism.) Nor does it involve lamenting the Absolute's absence. The Absolute occasions. It cannot be adequately thought in terms of the opposition between presence and absence.

5. The historicity of the question of style will be addressed in a more sustained manner in chapter 3. While it is possible to construct a number of different histories, here the direction will be given by the pamphlet written by Heinrich Hübsch in 1828, *In What Style Ought We to Build?* It is not as though the question of style did not exist before that date; rather, the argument is that Hübsch—at least within the context of architecture understood as a material practice—is concerned to locate style within a more general questioning of continuity.

6. Once conflict becomes constitutive of a period—even of a work—and once both style and the symbol are interarticulated with questions of historical time, then this brings an inevitable politics of time into play. Indeed, a politics of time will be essential to any understanding of modernity and the presence of the past within it. Here, of course, is a conception of conflict that is construed beyond the hold of reason's "cunning." Conflict is linked to action and thus to the structure of the decision, the latter being the marker of true conflict. This is not the conception of conflict that can posit a figure whose presence functions to resolve conflict by absorbing it. In Hegel, the prime example would be the role of the monarch in the *Philosophy of Right,* trans. T. M. Knox (Oxford: Clarendon, 1965). When

developing the position concerning the relationship between conflict and the state, Hegel argues that this conflict—conflict being the play of particulars without possible reconciliation—not only can be overcome by the "person of the monarch"; such a person (personality) is necessary in order that this threat to unity be overcome. The unity of the state and the unity of the monarch have an important affinity. In section 281 Hegel argues that

> in this unity lies the actual unity of the state, and it is only through this . . . that the unity of the state is saved from the risk of being drawn down into the sphere of particularity [*in die Sphäre der Besonderheit*] and its caprices, ends and opinions, and saved too from the war of factions [*dem Kampf der Faktionen*] around the throne and the enfeeblement and overthrow of the power of the state.

While this passage demands detailed comment, it is not difficult to note the reiteration of the problem of particularity, and its eventual reconciliation in an external figure that becomes its condition of possibility.

7. Equally, of course, attempting to counter one form of violence with another—where both are defined by ultimately illusory if nonetheless pragmatically real forms of an enforced conception of identity—becomes a way of understanding the clash between national forces even if one seeks legitimization within the domain of a posited international "good." I have attempted to undo this particular logic in terms of the way it structures the use of the term "terrorist" in contemporary politics. See "Judging Terrorism," *Parallax* 9, no. 1 (2003): 3–13.

8. The text in question is "Anmerkungen zum Oedipus," in J. C. F. Hölderlin, *Theoretische Schriften* (Hamburg: Felix Meiner, 1998), 94–95.

9. This point will be taken up in some detail at the end of chapter 1.

10. The key text in this context is Peter Osborne, *The Politics of Time* (London: Verso Books, 1993). While the project undertaken here differs in terms of how totalization would be thought, Osborne's pioneering work is the first to bring a series of philosophical projects together in terms of how differing temporalities of historical time are to be analyzed. The divergences become the ground of politics, even though the political possibilities are not reducible to each other. Also, of central importance for any thinking of time as having an inherently political dimension is Theodor Adorno's examination of the concept of progress. See "Progress" in his *Critical Models: Interventions and Catchwords,* trans. Henry W. Pickford (New York: Columbia University Press, 1998), 143–61. The nature of the relationship between this particular paper of Adorno's and Walter Benjamin's work is clear, as it is possible to chart a direct link between Benjamin's sustained critique of historicism and Adorno's critique of the temporality of progress.

11. The figures whose work gives direction to this book are Benjamin, Bataille, and Nietzsche. However, what follows is not meant as an act of scrupulous fidelity. Nonetheless, their work has served as a continual prompt. Benjamin's understanding of time, Bataille's formulation of materialism and his problematizing of

place, and Nietzsche's demythologizing refusal of the constraint of national thinking all play fundamental roles in the structure of the argument. And yet, to each of these possibilities a countermeasure can be found in the work in question. It is possible to argue against this project deploying Benjamin's utopianism, Bataille's theory of expenditure, and Nietzsche's thinking of time as eternal recurrence. This point is noted in advance. It is not as though there is an attempt to find tensions in these bodies of work. Rather, the project is always to find those moments which, in resisting ready assimilation to dominant traditions, demand modes of thinking whose particularity can be demonstrated. In this context that mode involves use.

12. "Theses on the Philosophy of History" was the title given to an English translation of this essay. However, "On the Concept of History" is the direct translation of the German title. Moreover, it is the title now used in the *Selected Writings,* and hence will be used in my own project here as well. The passage in question is the following: "Every image of the past that is not recognized by the present as one of its own concerns threatens to disappear irretrievably" (*SW* 4:391/*GS* 1.2:695).

13. In the context of this book, this is what will be described as myth. The projection becomes the projection of a future based on a yet-to-be-realized potential attributed to the past. The attribution constructs a unity that is then given a future. Such a move either constructs the past in a certain manner and thus establishes unity in terms of either place or ethnicity—or a combination of the two—and then projects that construction into the future; or the future becomes the place where the possibilities not realized in the present will occur. In both instances, the politics of myth involves the future. The struggle to enact modernity is therefore a struggle against this conception of myth; a conception that has its own disastrous politics. The basis for this formulation comes, in part, from Walter Benjamin's discussion of Ernst Jünger's *War and Warriors.* See "Theorien des deutschen Faschismus" in *SW* 2:312–21/*GS* 3:238–50.

14. The pragmatic symbol is the one that unifies different groups within a larger collective. It may be, of course, that the pragmatic is the truth of the symbol and that the aspiration for synthetic unity is the mythic or utopian dimension whose truth is its inherent falsity.

15. This point will return in chapter 5 in the consideration of Nietzsche's identification of "homelessness" and the modern. What is fundamental to the Nietzsche of *The Gay Science* is this identification. If there is a concern with style in the Nietzsche of this period, then it has to do with what occurs with style once it is no longer defined in national terms.

16. Moving beyond simple oppositions between the national and the international has to locate resistance elsewhere. Part of the project being worked out here is to identify what that "elsewhere" would involve. Rather than adopt a generalized conception of the global within which resistance would be unthinkable other than by acts of simple violence, it may be that complicating the nation-state by rethinking it in terms of regions may provide the way ahead. To the extent that this

is a real possibility, Homi Bhabha's working concept of the nation-state becomes valuable. (See his *Location of Culture* [London: Routledge, 1995].) Bhabha's work allows for the recognition of the necessity for a recasting of place—the place of the nation-state; this takes place by the inscription of the liminal as already inherent in the whole. It occurs, for example, in Bhabha's description of the nation-state in which the state's being is cast in the following terms:

> Once the liminality of the nation-state is established, and its difference is turned from the boundary "outside" to its finitude "within," the threat of cultural difference is no longer a problem of the other people. It becomes a question of the otherness of the people-as-one. (301)

This project will demand how what is identified as "the otherness of the people-as-one" can be given another description. In other words, how are identity and difference, as present in this setup, to be thought? What is the mode of being proper to this conception of the nation-state and thus of an already present otherness? The answer to be developed in the argument to come is in terms of the cosmopolitan.

17. The engagement with Frampton's work will be with his papers "Towards a Critical Regionalism: Six Points for an Architecture of Resistance" and "Rappel à l'Ordre: The Case for the Tectonic." Both are now collected in his *Labour, Work and Architecture: Collected Essays on Architecture and Design* (London: Phaidon, 2002).

Chapter 1

1. In conjunction with Beatrice Hansen, I have brought together a number of texts in order to offer a sustained investigation of Walter Benjamin's work on Romanticism. See Beatrice Hansen and Andrew Benjamin, eds., *Walter Benjamin and Romanticism* (London: Continuum Books, 2002).

2. For an important discussion of Benjamin's work that pays attention to his relationship to Romanticism, see Howard Caygill, *Walter Benjamin: The Color of Experience* (London: Routledge, 1998).

3. Even though it is not given a direct thematic formulation, fundamental to this project is the insistence on what could be described as Benjamin's formalism. Two instances of this can be cited. The first is his argument for breaking the link between the caesura and a theory of tragedy. The caesura cannot be reduced to its place in such a theory. The importance of this move is that the caesura acquires a formal quality not determined by a specific content. As such, the genealogy of the term can be noted, and its productive potential can be released. (Benjamin's formalism gives the caesura its "afterlife.") The second instance—which will be taken up in chapters 2 and 3—is Benjamin's use of the term "structure." This occurs most significantly in *The Arcades Project* in which he argues that a genuine interruption,

which he terms "the liberation from an epoch," has the "structure of an awakening" (G1, 7). What this means is that centrality is attributed to the structure, not to a specific content. The current fetishization of content—and thus of the political dimension of art, for example, being located in subject matter—works with the nostalgic belief in the eventual arrival of the correct image. Benjamin's formalism—and, it could be argued, his productive amoralism—was the insistence of structure.

4. While not defining its presence in relation to the Absolute, Carol Jacobs indicates the extension of "criticism" in her treatment of the relationship between criticism and translation. See *In the Language of Walter Benjamin* (Baltimore: Johns Hopkins University Press, 1999), 286–87.

5. I have tried to develop this argument in my *Present Hope* (London: Routledge, 1997).

6. The full quotation is: "It could be that the continuity of tradition is only an appearance. But if this is the case, then it is precisely the persistence of this appearance of permanence that establishes continuity of appearance."

7. This is a complex and perhaps difficult claim. The argument is straightforward, however. Forms of continuity, rather than being either natural or inevitable, are forms of time. Time is given such forms for specific ends. Interrupting such concepts of time works, first, to restrict the realization of those ends; and second, to show that such concepts are neither natural nor inevitable; and finally it shows that time is a site of contestation. The conflict between continuity and discontinuity is the truth of time. Interruption, precisely because it reveals the work of construction, stages time's truth.

8. I have developed an argument for the relevance of Schlegel as providing the means for a reconsideration of the Absolute in my *Philosophy's Literature* (Manchester: Clinamen, 2001).

9. Immanuel Kant, "Idea for a Universal History with a Cosmopolitan Purpose," in *Kant's Political Writings,* ed. H. Reiss (Cambridge: Cambridge University Press, 1977), 41–53; Immanuel Kant, "Idee zu einer allgemeinen Geschichte in weltbürgerlicher Absicht," in *Kants Werke Akademie-Textausgabe,* vol. 8 (Berlin: Walter de Gruyter, 1968), 15–32. (The passage cited is on pages 44–45 and 21 of these English and German versions, respectively.) The position to be developed throughout this book—contrary to Kant's use of a German equivalent—is that the cosmopolitan is the recognition of dissensus as yielding sites of negotiation. This is a setup which in general is thought beyond the hold of a linear conception of time. Allowing for such a position necessitates another politics of time.

Chapter 2

1. The most sustained of recent attempts to treat the question of fashion in Benjamin's writings is Peter Wollen, "The Concept of Fashion in The Arcades

Project," *Boundary 2,* vol. 30, no. 1 (Spring 2003). While this is an excellent paper, it does not pursue the question of fashion as a temporal marker. Fashion becomes identified—almost exclusively—with its literal presence.

2. The passage in question has to do with the consequences of revolutionary art. The "theses" pertaining to such an art form, for Benjamin, "brush aside a number of outmoded concepts such as creativity and genius, eternal value and mystery" (*SW* 4:252/*GS* 1.2:473). While the passage goes on to show how there is a link between the retention of such concepts and the goals of fascism, it is also the case that "eternal value" is identified as an enduring concept.

3. While the argument to be advanced in this chapter concerns the tension between Benjamin's politics of time and the identification of the utopian in history, Benjamin's actual commitment to utopianism can be questioned in its own right. For an important contribution to this questioning, see Detlef Mertins, "Walter Benjamin and the Tectonic Unconscious," in *Walter Benjamin and Art,* ed. Andrew Benjamin (London: Continuum, 2005), 148–63.

4. The literature on Benjamin's "On the Concept of History" ("Theses on the Philosophy of History") is extensive. Due to the restrictions of space, there cannot be a sustained engagement with that literature. It is also the case that because what is central to this project is developing the implicit politics of time in Benjamin's argument via a specific engagement with the temporality of fashion, attention has to be paid to identifying the nature of the distinction between the utopian impulse within that work and the temporalization of the political. Important contributions in this area can be found in *Walter Benjamin and History,* ed. Andrew Benjamin (London: Continuum, 2005); for instance, see Werner Hammacher's "Now: Walter Benjamin on Historical Time." See also Eduardo Cadava, *Words of Light: Theses on the Photography of History* (Princeton, N.J.: Princeton University Press, 1997); and Fritz Breithaupt, "History as the Delayed Disintegration of Phenomena," in *Benjamin's Ghosts: Interventions in Contemporary Literary and Cultural Theory,* ed. Gerhard Richter (Stanford, Calif.: Stanford University Press, 2002), 191–204.

5. The translation by Zohn has been slightly modified. The modifications, which perhaps result in a less elegant version, have been made in order to stick more closely to Benjamin's vocabulary. Moreover, these changes should allow the vocabulary of Thesis XIV to have a greater affinity with the presence of similar terms in other texts by Benjamin.

6. The term "constructivist" has been used because it follows from Benjamin's own description of history as a "construction." It is not intended to have further extension. Once an insistence is made upon the activity of construction, it then becomes possible to connect it to strategy. The term joining the two is "cunning." This is the position that will be taken up via a consideration of G1, 7 from *The Arcades Project.*

7. I am not going to argue that Benjamin's response to Nietzsche is necessarily correct. It may be that while he misconstrues Nietzsche's own position, he has

developed a sustained critique of a different version of historicism. For a spirited defense of Nietzsche on "eternal return," see David Farrell Krell, *Infectious Nietzsche* (Bloomington: Indiana University Press, 1996), 56–82. Krell's response is doubly interesting, since while he does not engage with Benjamin directly, he does take up aspects of Karl Löwith's critique, as does Benjamin.

8. This term "urgeschichtlichen" is left untranslated in the English edition of *The Arcades Project.* Within the corpus of Benjamin's own work, the use of this term—or more accurately, the use of the prefix "ur"—is interesting, since it recalls his critique of Goethe's conception of art. In Benjamin's terms, Goethe conceived of art as having its ground in an "ideal." Furthermore, "in relation to the ideal, the single work remains, as it were a torso. It is an individual endeavor to represent the archetype [*das Urbild darzustellen*]" (*SW* 1:180/*GS* 1.1:114). This relation to the "Urbild" made the work uncriticizable and exposed the relation that Goethe's concept of art had to myth and to the structure of mythological time. However, Benjamin admits a different function of the "ur" prefix as manifest in the notion of the origin in *The Origin of German Tragic Drama,* as is evidenced, for instance, in its prologue (*OT* 27–56/*GS* 1.1:207–37). The Benjaminian notion of the *Ursprung* will be further discussed at the end of the present chapter.

9. "Cunning" becomes a term designating forms of political action: actions taken by all sides in the political divide. For Benjamin this is a position involving fashion. See *The Arcades Project* E5a, 5: "The mighty [*Die Herrschenden*] seek to secure their position with blood (police), with cunning (fashion), with magic (pomp)."

10. This is not simply a linguistic point. Benjamin must have been acutely aware of the use of "Einst und Jetzt" displays in German department stores—specifically Berlin—in the late 1920s. In a thesis whose concern is the temporality of fashion, such a link must be retained. Hence, the double register of fashion. It allows the force of the distancing of the temporality of fashion to take on a greater intensity. In regard to this form of display, see Janet Ward, *Weimar Surfaces: Urban Visual Culture in 1920s Weimar* (Berkeley: University of California Press, 2001), 217.

11. In regard to this "blasphemy," see Thesis B (*SW* 4:397/*GS* 1.2:704).

12. Perhaps, contrary to certain expectations by Benjamin, I am insisting on a non-theological reading of theology. Theology becomes a way of thinking the time of interruption. This is not the politicization of the theological; it is the repositioning of the theological as a philosophy of historical time. The best overall study of the relation between theology and politics in Benjamin is Tiedemann's early essay. While not identifying theology simply with time, he points the way toward stripping theology of its being the language of God and its becoming the temporality of interruption occurring in the name of hope. See Rolf Tiedemann, "Historical Materialism or Political Messianism? An Interpretation of the Theses 'On the Concept of History,'" in *Benjamin: Philosophy, Aesthetics, History,* ed. Gary Smith (Chicago: University of Chicago Press, 1989), 175–210. See also Howard

Caygill, "Non-Messianic Political Theology in Benjamin's 'Concept of History,'" in *Walter Benjamin and History,* ed. Andrew Benjamin. I have also tried to explore these topics throughout my *Present Hope* (London: Routledge, 1997).

Chapter 3

1. Central to this project is the argument that the style debate in Germany provides a locus in which to specifically discuss the emergence of modern architecture, and more generally modernity as articulated within a founding interruption. Both possibilities are connected to the necessary severance of style and appearance. This severance is the implicit generator of the debate. The majority of the texts comprising the debate can be found in English in W. Herrmann, ed., *In What Style Shall We Build? The German Debate on Architectural Style* (Santa Monica, Calif.: Getty Center, 1992).

2. Walter Curt Behrendt, *Der Sieg des neuen Baustils* (Stuttgart: Wedekind, 1927). This book is available in English as *The Victory of the New Building Style,* trans. Harry Francis Mallgrave (Los Angeles: Getty Research Institute for the History of Art and the Humanities, 2000).

3. This text is available in English translation in the anthology edited by W. Herrmann, and cited in note 1 for this chapter. The original German edition is H. Hübsch, *In welchem Style sollen wir bauen?* (Karlsruhe, 1828).

4. Jacques Ignace Hittorff, *L'architecture antique de la Sicile.* This text is of enormous historical importance in relation to the debate concerning polychromatic antiquities that occurred in the mid-nineteenth century in Europe. While the debate was on one level explicable purely in architectural terms, it is also the case that it involved a problem of Europe's identity in terms of a type of purity. To this end, Nietzsche's description in *The Gay Science* of the "modern" person as "racially mixed" from the start has to be read in conjunction with this anxiety about purity and color. I have taken up this aspect of Nietzsche's work in chapter 5. (References to Nietzsche are provided in chapter 4, note 7.) For an overview of the debate concerning polychromatic antiquities set within a historical context, see David Van Zanten, *The Architectural Polychromy of the 1830s* (New York: Garland, 1977). In regard to the question of whiteness and this debate, see Mark Wigely, *White Walls* (Cambridge, Mass.: MIT Press, 1995).

5. J. W. von Goethe, *Werke* (Hamburger edition), vol. 12, *Schriften zur Kunst* (Munich: C. H. Beck, 1994), 176.

6. References to Winckelmann are to the following editions. J. Winckelmann, *Reflections on the Painting and Sculpture of the Greeks,* trans. Henry Fuseli (London: Routledge, 1999). (This edition is a facsimile reproduction of the 1765 edition.) J. Winckelmann, *Kleine Schriften zur Geschichte der Kunst des Altertums,* vol. 1 (Leipzig: Insel, 1925). References in text to these English and German

versions are abbreviated as *R* and *KS,* respectively. For an overview of the role of imitation in Winckelmann, one that demonstrates the importance of Raphael, see Michael Fried, "Antiquity Now: Reading Winckelmann on Imitation," *October* 37 (Summer 1986): 87–92. For an extended discussion of the general question of the relationship between German thought and antiquity—the Greek world—see David Ferris, *The Silent Urns* (Stanford, Calif.: Stanford University Press, 2000).

7. Winckelmann is, of course, not alone in defining a relation to the Greeks in terms of imitation. Sir Joshua Reynolds (1774) also defines a relation to the past in these terms. See in this regard the discussion of "Genius" in Discourse VI (Joshua Reynolds, *Discourses on Art,* ed. Robert R. Wark [New Haven, Conn.: Yale University Press, 1975]). What is interesting in comparing Reynolds and Winckelmann is that for the former there is no sense of an "ideal." Hence imitation is inextricably bound up with the problem of copying. In this regard, Reynolds argues that

> When I speak of the habitual imitation and continued study of the masters, it is not to be understood that I advise any endeavour to copy the exact peculiar colour and complexion of another man's mind: the success of such an attempt must be always like his, who imitates exactly the air, manners and gestures, of him who he admires. His model may be excellent, but the copy will be ridiculous; this ridicule does not arise from his having imitated, but from his not having chose the right mode of imitation. (100)

If there is a counter to Reynolds's work in the same way that the following interpretation of Hübsch is the counter to Winckelmann, then it is to be found in Hazlitt's critique of Reynolds. See William Hazlitt, *Complete Works,* ed. P. P. Howe (New York: Aros, 1967), 14:64–70.

8. For a discussion of the role played by Raphael and the use made of his work in the history of philosophy, see the account given by Wolfgang von Löhneysen in *Raffel unter den Philosophen: Philosophen über Raffel* (Berlin: Duncker & Humbolt, 1992). While Winckelmann is treated throughout, this point is taken up on 40–41.

9. See Winckelmann's treatment of Michelangelo in the *Reflections* and then a subsequent mention where he is discussed in relation to Borromini: Johann Joachim Winckelmann, "Anmerkungen über die Baukunst der Alten," in *Schriften und Nachlaß,* vol. 3, *Schriften zur antiken Baukunst* (Mainz: Philipp von Zabern, 2001), 61.

10. For a discussion of the etymological meaning of this term in Longinus, see D. A. Russell, *Longinus' "On the Sublime"* (Oxford: Clarendon, 1964), 74–75.

11. Schelling comments on this passage from Winckelmann. He argues that as an "expression of the soul" *(Ausdruck der Seele)* it is not drawn from experience. What is expressed, therefore, is an "idea" *(Idee)* that transcends nature. Schelling confirms the inherent idealism of Winckelmann, though he also makes it clear that paintings can "express" ideals. See F. W. J. Schelling, "Philosophie der Kunst,"

in *Ausgewählte Schriften,* vol. 2 (Frankfurt am Main: Suhrkamp, 1985), 385–86. Schelling also discusses the importance of Raphael. Clearly, it is an examination that draws heavily on Winckelmann's original analysis.

12. The most informative account of Hübsch's importance within German architectural theory is Barry Bergdoll, "Archaeology vs. History: Heinrich Hübsch's Critique of Neoclassicism and the Beginnings of Historicism in German Architectural Theory," *Oxford Art Journal* 5, no. 2 (1983): 3–13.

13. Herrmann, ed., *In What Style,* 65; Hübsch, *In welchem Style,* 1.

14. Karl Boetticher, "Das Prinzip der Hellenischen und Germanischen Bauweise hinsichtlich der Übertragung in die Bauweise unserer Tage," in *Schinkel zu Ehren,* ed. Julius Posner (Herausgegeben vom Architekten—und Ingenieur-Verein zu Berlin, 1978), 22. The text can be found in English in Herrmann, ed., *In What Style.* The best introduction to Boetticher's work is Mitchell Schwarzer, "Ontology and Representation in Karl Boetticher's Theory of Tectonics," *Journal of the Society of Architectural Historians,* no. 52 (September 1993): 267–380. See also Frampton, "Rappel à l'Ordre."

15. While Hübsch will resolve the problem of style by recourse to Rundbogenstil, it remains the case that the tension introduced by the question at the end of his pamphlet's title endures. The straightforward point argued here is that the very presence of this mark signals the severance of style and appearance. It can, of course, be noted elsewhere. However, there is great force in the proposition that once style becomes a question what that means is that the claim that there is a given or singular appearance proper to a style is contestable. That contestability is again the sign of the severance. It should not be thought, however, that there is only one way through this severance. In a text of the same name as that of Hübsch's pamphlet, Carl Albert Rosenthal argued, in 1848, that the break with the classicist tradition had to be undertaken in the name of the unique nature of the people; more exactly, the "soul" of the people. The point remains, however, that severance is announced and thus a type of discontinuity is affirmed. The very fact that its results or that which grounds it lacks unanimity only reinforces the general point that with the severance, what emerges is the problem of appearance. It is problematic precisely because the appearance of appearance is no longer determined in advance. See Carl Albert Rosenthal, "In welchem Style sollen wir bauen?" *Zeitschrift für praktischen Baukunst* 4 (1844).

16. Karl Boetticher, *Die Tektonik der Hellenen* (Potsdam: Riegel, 1852), 20. Here Boetticher distinguishes between *Werkform* and *Kunstform.*

17. While a definition of *Werkform* is relatively unproblematic, since it refers to the materiality of the object and thus is connected to the object's tectonic presence, the term *Kunstform* presents far greater problems. Harry Francis Mallgrave, in his introduction to the English translation of Semper's *The Four Elements of Architecture* (trans. Harry Francis Mallgrave and Wolfgang Herman [Cambridge: Cambridge University Press, 1989]), defines *Kunstform* as "a conceptual veil that overlays the column giving it its characteristic expression" (37). What is essential to *Kunstform* is that it is the site of an appearance that is not reducible to mere

tectonic presence. The force is the conceptual element. It is this aspect that allows for a link between the *Kunstform* and the public. *Kunstform* is not just appearance; it is an appearance whose content is recognized.

18. Peter Behrens, "Kunst und Technik," 279.

19. Henry van de Velde, *Déblaiement d'art* (Paris, 1894), 15.

20. What makes this passage of such crucial importance for this project is its formulation of what was identified in the introduction as Benjamin's formalism. In this context, it can also be read as an implicit critique of the explicit violence inherent in van de Velde's conception of destruction. Again this indicates the fundamental difference between Benjamin and the differing permutation of nihilism within European thought in this period.

21. For an important development of this point, see Stanley Corngold, "Obscurity Shadows the Semblance Whose Obliteration Promises Redemption: Reflexions on Benjamin's 'Goethe's Elective Affinities,'" in *Benjamin's Ghosts,* ed. Gerhard Richter, especially 155–56.

Chapter 4

1. There are at least two texts by Bataille that are of fundamental importance for an understanding of what he means by "materialism." The first is the text cited in note 2 of the introduction and the second is another early essay, "Le bas matérialisme et la gnose," in *Oeuvres complètes,* 1:220–27. Both of these short texts by Bataille will reappear in the extended discussion of Bataille's interpretation of Manet in chapter 7. For all its problems, the most sustained account of Bataille's engagement with architecture is D. Hollier, *Against Architecture* (Cambridge, Mass.: MIT Press, 1989).

2. While on one level there are few affinities between the work of Adorno and Bataille, there is an interesting connection concerning the detail of materialism. For Adorno, the simple positing of materialism as a counter to idealism was not philosophically sufficient. Hence, his argument that the position for and against metaphysics is itself metaphysical. The development of a theory of materialism occurring through the lectures now published as *Metaphysics: Concept and Problems* (Stanford, Calif.: Stanford University Press, 2001) details the difficulties of establishing materialism in contradistinction to an idealist conception of materialism. Part of the force of Bataille's position is that not only does it have registers elsewhere in the history of materialism, but it shows that there is a conflict over the nature of matter at the heart of differing philosophies of materialism. Matter is never just matter.

3. I have dealt with this distinction in considerable detail in my *Architectural Philosophy* (London: Continuum, 2000).

4. The obvious corollary here is with myth. The argument has to be that the process of demythologization cannot itself be mythic. Undoing the work of myth

cannot involve the promulgation of a counter-myth. Undoing myth necessitates showing what is at work as mythic.

5. Adolf Loos, "Ornament and Crime," in *The Architecture of Adolf Loos* (London: Arts Council of Great Britain, 1987), 102.

6. Again, it has to be noted that this position is more complicated than this formulation notes. Clearly, there are at least two subject positions; the subject of historicism and the subject position in which the caesura identifies a point of affirmation. It should not be thought that these two subject positions of necessity have to identify two discrete individuals. The fact that a given individual could hold to both does not reflect a logical contradiction, it simply identifies the founding tensions within the fabric of modernity.

7. Friedrich Nietzsche, *Die fröhliche Wissenschaft,* in *Kritische Studienausgabe,* vol. 3 (Berlin: Walter de Gruyter, 1999), 310; Friedrich Nietzsche, *The Gay Science,* trans. Walter Kaufmann (New York: Vintage Books, 1974), 340.

8. Nietzsche, *Die fröhliche Wissenschaft,* 313.

9. What is being suggested here is that subjectivity, in being the site of alterity—the reality that the self is both itself and other to itself, this being the reality of the distinction between the unconscious and the conscious—can be understood as the locus of what I will describe as an original difference. (I have developed this term in a number of places. See, for example, *The Plural Event* [London: Routledge, 1993].) This is, of course, the founding insight of psychoanalysis. An important investigation of the way the unconscious can be understood as a form of alterity can be found in the work of Jean Laplanche. See in particular his "Le temps et l'autre" in *La revolution copernicienne inachevée* (Paris: Aubier, 1992), 359–85; and "Court traité de l'inconscient," in *Entre seduction et inspiration: L'homme* (Paris: Presses Universitaires de France, 1999), 67–115.

10. In this regard, I would agree with Esra Akcan's argument that "place needs to be an issue not because we search for the essence and unchanging nature of locales, but because there is an ideological, economic and ecological need for a non-essentialist, unbounded and cosmopolitan conception of place in the age of global capitalism." See Esra Akcan, "Critical Practice in the Global Era: The Question concerning Other Geographies," *Architectural Theory Review* 7, no 1 (2002): 53.

11. As concerns the work done by philosophers, there is now an extensive body of material on the question of place. Two of the most important thinkers in this domain are Edward Casey and Jeff Malpas. In regard to the former, see *Getting Back into Place* (Bloomington: Indiana University Press, 1993); *The Fate of Place* (Los Angeles: University of California Press, 1997); and *Representing Place* (Minneapolis: University of Minnesota Press, 2002). In regard to the latter, see *Place and Experience: A Philosophical Topography* (Cambridge: Cambridge University Press, 1999).

12. One the most exacting studies of Heidegger's understanding of the "earth" can be found in Michel Haar, *Chant de la terre* (Paris: L'Herner, 1985).

13. Frampton, "Rappel à l'Ordre," 81. Frampton has recently clarified and expanded the basis of a number of his arguments during the course of a long published conversation with Stan Allen and Hal Foster. See Stan Allen and Hal Foster, "A Conversation with Kenneth Frampton," *October* 106 (Fall 2003): 35–59.

14. Frampton, "Rappel à l'Ordre," 85.

15. Ibid., 89.

16. Martin Heidegger, *Basic Writings,* ed. and trans. David Farrell Krell (San Francisco: HarperCollins, 1993), 363. The original German text is in *Vorträge und Aufsätze* (Tübingen: Neske, 1990), 156.

Chapter 5

1. Consistent with the discussion of Kant's conception of cosmopolitanism and its relation to an explicit conception of historical time—see chapter 1—the point can be made again that time and place are already interarticulated. Here the argument is that the possibility of the cosmopolitan lies in the complex nature of the present. For Kant, as has already been argued, there is a single—and thus continuous—line leading to the future. Everything is distributed along that line. Place and a linear temporal narrative are articulated together. While secular in orientation precisely because it deploys an understanding of development as driven by the continuity of engagement with the dictates of reason, the inherent temporality of development is a continuous linear response to the founding imperatives of nature. In the Eighth Proposition of the already cited paper by Kant on cosmopolitanism, he argues that despite the inadequacies of the present—identified as that which occurs "now" *(jetzt)*—it remains the case that a "feeling" for unity exists and thus this gives rise to the "hope" *(Hoffnung)* that transformations will occur leading to the attainment of "the highest purpose of nature, a universal cosmopolitan existence" (Kant, "Idea for a Universal History," 50/Kant, "Idee zu einer allgemeinen Geschichte," 28). The argument against this position is not pitched against accord per se, though it should be added that the cosmopolitan as a state to be attained needs to be rethought not as an end but as an indeterminate universal whose content is the site of continual negotiation. The difficulty, as has already been argued, is the inherent teleology in the formulation of the position.

2. This possibility, which is situated in the work of both Marx and Henri Pirenne, is developed in slightly greater detail in chapter 6. It would be interesting to pursue this point. Another possibility that could also be explored would arise from developing an urban diagram based, initially, on the relationship between Marx described as the movement of "merchant capital" and the flows of information allowed by the international and the now digitally created movements of capital. (The latter while connected are not the same.)

3. Herrmann, ed., *In What Style,* 52.

4. There are a number of well-researched and documented general books on Plečnik. To this end, see Damjan Prelovsek, *Jože Plečnik: 1872–1957: Architectura Perennis* (New Haven, Conn.: Yale University Press, 1997); and Peter Krecic, *Plečnik: The Complete Works* (New York: Whitney Library of Design, 1993). The task of this engagement with Plečnik is not to offer a sustained interpretation of the work as a whole. The project is more strategic: the relationship between the use of symbols and what will be identified as a projective sense of national identity. For that reason, the analysis is limited to the Three Bridges and the National University Library.

5. D. Grabijan, *Plečnik in nejova sola* (Maribor, 1968), 114, quoted in Prelovsek, *Jože Plečnik,* 13. Prelovsek, again quoting from an unpublished letter, draws an important connection between Alois Riegel's conception of *Kunstwollen* and this position of Plečnik 's. Even though it is beyond the scope of this chapter to engage, in any sustained way, with the question raised by that connection, it is nonetheless the case that while there are affinities, a moment of doubt can be introduced once it is recognized that for Plečnik the "inner nerve of art" allows what was initially inchoate to cohere. Hence random symbols are organized by an external idea. It is not straightforward that this capacity to unify that which is—at least on the level of appearance—without unity is what Riegel had in mind.

6. Reproduced in Prelovsek, *Jože Plečnik,* 255.

7. It is not as though this is a contested point. The question is how to interpret the presence of the Mediterranean. For example, Peter Krecic argues the following in his "Jože Plečnik and Art Deco," trans. Marjan Golobic, *Journal of Decorative and Propaganda Arts,* no. 17 (Fall 1990), 26–36.

> When Plečnik returned to Ljubljana, he tried to continue the Mediterranean plasticity, which appeared to be the most efficient means of imparting a monumental aura to the city. His bridges spanning the Ljubljanica, especially Tromostovje, are reminiscent of Venice. The National and University Library is a monumental cube-shaped building, a true Italian palazzo featuring a certain feel for the symbolism of its parts. (35)

In regard to this palazzo, Prelovsek suggests that its model is more precise. He proposes that the "architect liked to stress that the corner of his library on the New Square radiated a monumentality similar to that of the Palazzo Medici-Riccardi in Florence" (*Jože Plečnik,* 256). His source is Edo Ravnika.

8. Robert Gilkey Dyck makes a similar argument in relation to Plečnik's urban projects in general: "His idea was to improve the 'legibility' of the city by appealing to the ability of people to read and identify with traditional forms, in this way linking the city enduringly to its Mediterranean origins" (15). See his "Classical Antecedents of Plečnik's Urban Design for Ljubljana," *The Classicist,* no. 5 (1998–99): 11–17. Dyck goes on to argue that the design for the Central Market, in addition to the other elements of the city, took the Greek city as their model.

9. In the Slovenian context, nationalism has more than one dimension. Indeed, Plečnik's nationalism is the more complex precisely because it is projective. At more or less the same time, the nationalist architecture of Janez Jager and Ivan Vurnik, for example, was Pan-Slavic in orientation. It was articulated through the use of a different set of symbols and form of decoration. See, for example, Vurnik's exterior to the Mutual Economic Bank (1921). These different possibilities are touched on by Stane Bernik in his article "Slovene Architecture from Secession to Expression to Functionalism," trans. Marjan Golobic, *Journal of Decorative and Propaganda Arts,* no. 17 (Fall 1990): 42–54.

10. It is interesting to note in this regard the popularity of Plečnik among postmodern architectural theory. What that celebration of his work consistently overlooked was the organizational force of a projective sense of nationalism determined at every moment by an overarching idea. It delimited the work. As a countermeasure, the 1986 exhibition of Plečnik's work in Paris was defined by the curator as a project that aims as much at resisting that attempt to assimilate the architecture of Plečnik to the concerns of postmodernism as to establish the importance of his work in its own right. This had to do with identifying a specific series of ethical concerns within the work. In addition, part of the presentation of the work was in terms of its precluding the international in the name of the national and thus to a certain extent the regional. For a discussion of these points, see the essay "Modern or PostModern: A Question of Ethics?" by François Burkhardt in the exhibition catalog: François Burkhardt, Claude Eveno, and Boris Podrecca, eds., *Jože Plečnik: Architect: 1872–1957,* trans. Carol Volk (Cambridge, Mass.: MIT Press, 1989), 108–19. (This is the English translation of the original French catalog.) For a more informal—though slightly more polemical—version, see the interview given by Burkhardt at the time of the exhibition: François Burkhardt and Alain Pélissier, "Jože Plečnik entre tradition et modernité," *Techniques & Architecture,* no. 365 (1986): 25–27.

11. The argument here is limited to postmodern architecture. The philosophical position that is often identified with postmodernism, outlined in Jean-François Lyotard's *La condition postmodern* (Paris: Éditions de Minuit, 1979), differs radically both in approach and political orientation.

12. John Macarthur, "Australian Baroque," *Architecture Australia,* March–April 2001, 52–58.

13. The recently published volume *Surface Consciousness,* ed. Mark Taylor (London: Wiley, 2003), is one of the most important anthologies of work in this area. Of particular interest in this regard is the contribution by Michael J. Ostwald that treats, in part, work by Lyons. See his "Seduction, Subversion and Predation: Surface Characteristics," 75–81.

14. The work of Herzog and de Meuron has attracted increasingly more sophisticated criticism. One of the most important is a recent article by William J. R. Curtis, "'Enigmas of Surface and Depth': The Architecture of Herzog and

de Meuron," *El Croquis* 84 (1997): 32–49. Not only is Curtis right to see the work as linked to a heightening of "our perception of the world"; he shows in what way this is an effect of the building's construction. "Free facades" do not occur by chance. In contrast is Jeffrey Kipnis, who, while rightly calling attention to questions of perception, can account for the work only in terms of effect. As such, the architecture loses its architectural quality and becomes no more than a simulacrum. (See Kipnis, "The Cunning of Cosmetics," *El Croquis* 84 [1997]: 22–28.)

Chapter 6

1. This paper was given as the Martineau Lecture in Launceston and Hobart, Australia, on September 24 and 26, 2002, respectively. I would like to thank the generosity of the Department of Philosophy at the University of Tasmania and in particular Professor Jeffrey Malpas for inviting me to give the lecture. While the argument has been truncated in order to conform to the constraints of a lecture, I have used the notes to indicate additional arguments, and to provide some of the background to the positions presented in the lecture itself.

2. The reference here in the Australian context is to the policy of the current government that maintains a position of "mandatory detention" for all arriving refugees. The current official opposition party, namely the Australian Labour Party, unanimously supports this position. The only sustained voice of dissent comes from the Australian Greens.

3. Andrew Lokan, "From Recognition to Reconciliation: The Function of Aboriginal Rights Law," *Melbourne University Law Review* 3, no. 23 (1999): 8. I have drawn on this article as it provides an argument for the extension of pre-contact Aboriginal law to spheres other than those delimited by property. However, the subsequent claims made in the lecture concerning the effect of this argument on the "subject of right" or claims about movement and fixity should not be attributed to Lokan. His article is, however, of fundamental importance in order to see the full impact, from within a strictly legal frame of reference, of giving a binding status to Aboriginal law. For an important study of the consequences of the Mabo decision for both legal and political philosophy, see Paul Patton, "The Translation of Indigenous Land into Property: The Mere Analogy of English Jurisprudence," *Parallax* 6, no. 1 (2000): 25–38.

4. While it cannot be pursued in this context, the philosophical reference is to two fragments attributed to Heraclitus, fragments 114 and 44. The opening line of the first is central here:

> ξὺν νόῳ λέγοντας ἰσχυρίζεσθαι χρὴ τῷ ξυνῷ πάντων, ὅκωσπερ νόμῳ πόλις

Speaking with understanding, they must hold to what is shared by all, as a city holds to its law.

Fragment 44 reads in its entirety:

μάχεσθαι χρὴ τὸν δῆμον ὑπερ τοῦ νόμου ὅκως ὑπὲρ τείχεος

The people must fight for the law as for their city wall.

These fragments taken together present a complex central to any understanding of place. Taken together they join human commonality, defined by the shared; law (*nomos*), though not a statute but as the ineliminability of normativity; and the "city wall" as that which discloses the space of normativity and thus of commonality. Normativity is not to be identified with the content of any particular social formation. Normativity becomes a name for the transcendental condition for sociality. However, that condition is not realized in the abstract; it is always located. The location is given by the "city wall." Once the wall forms an inextricable part of the interplay of the common and *nomos,* then the conditions of entry and exit—the right to enter and the right to leave—are themselves part of the same setup. The arresting of movement that establishes the wall re-creates movement back and forth through the wall as an inherent part of human being. What this suggests is that the refugee is not the exception but forms part of the structure of human sociality once the latter takes its point of departure from the interplay of commonality, normativity, and place.

5. This should not be read as a claim about nomads where such claims privilege certain anthropological groups over others. It is an argument concerning priority, both temporal and evaluative. Once settlement and thus stasis are seen as moments of arrest such that what characterizes activity is movement, then the static indicates where lines of movement have become dense by beginning to cohere and interact. An instance of this position is Henri Pirenne's argument that the origins of towns are to be understood in terms of the intersection of roads. (See his *Les villes du moyen-age* [Paris: Presses Universitaires de France, 1978].) There is a sense in which this can be viewed as an extension of Marx's argument that what flows along these roads are not goods in any straightforward sense but "merchant capital." (See the discussion of the origin of "merchant capital" in Marx's *Capital,* vol. 2.)

In addition to this work, there is a great deal of historical evidence that indicates the way the history of the frontier is to be defined in terms of movement. The place and thus the fixity of the frontier then become defined in terms of permeability and negotiation. Even the refusal of entry is a trope of movement rather than the other way around. The following works begin to open up ways of taking movement as fundamental to settlement. I am not suggesting that the following historians define their work in that way. The argument is that the way porosity and negotiation define borders and frontiers allows for the reinscription of movement as fundamental to place.

See Angus Mackay, "Religion, Culture and Ideology in the Late Medieval Castilian-Granadan Frontier," in *Medieval Frontier Societies* (Oxford: Clarendon, 1989), 217–43; C. R. Whittaker, *Frontiers of the Roman Empire* (Baltimore: Johns Hopkins University Press, 1994), especially 98–132; and Emilio Sereni, *Paseaggio Agrario Italiano* (Rome: Editori Laterza, 1999). The "porous" now as a condition of the city rather than simply the border or frontier is used by Massimo Cacciari in his discussion of Naples. See his "Naples the Porous City," *Alphabet City* 6 (1998): 28–44. There is an important point here concerning the relationship between terminology and a study of urban conditions. Once an urban setting is seen as a series of frontiers and borders, then the movement through urban conditions can be understood as a series of negotiations with more or less porous sections. Fundamental to the analysis is not the presence of literal borders or frontiers, but of a way of reconfiguring the site in those terms.

6. Walter Benjamin, "Uber der Begriff der Geschichte," in *GS* 1.2:696.

7. The classical formulation of this position is in Adorno and Horkheimer's *Dialektik der Aufklärung*, published in Theodor W. Adorno, *Gesammelte Schriften*, vol. 3 (Frankfurt am Main: Suhrkamp, 1997).

8. This formulation—or similar ones—occurs a number of times in Aristotle's writings. In regard to the *Politics*, the two of central importance are 1253a19 and 1278b20–21. It is the second of these passages that is the most significant, as it also involves a discussion of life. There are two elements of this difficult section of the text that demand comment. The first is that communal existence is that which allows the individual—and here it has to be the individual as an instance of the political animal—to realize "the good life." The use of the term "good" has to be understood in terms of that which is proper to human being, where the latter is understood as a political animal. The second is that while Aristotle will use the term "*bios*" to isolate "mere life," he claims that even in that form of life there is the ineliminable presence of that which is essential to human being. This is consistent with the earlier formulation in the *Politics* in which the possibility of being "*apolis*" is found only in gods or those without humanity. While more argumentation needs to be adduced, these references indicate the basis on which it is possible to argue that human animality is indistinguishable from human sociality.

9. Bhabha, *Location of Culture*, 175.

10. These questions set the frame of reference in which it becomes possible to repeat the engagement with Ashton Raggatt MacDougall's National Museum of Australia built in 2001. In sum, the argument already deployed involves the following considerations. The position advanced here can be easily differentiated from any easy identification with arguments stemming from what could be described as a "postmodern" history of (symbolic) architectural fragmentation. Postmodernism, which is best understood as a moment within the history of ornament rather than anything involving a temporal disjunction with the present, operates by deploying decontextualized symbols. The form of decontextualization involved

uses "dead" symbols. Any act of recall is superfluous. Symbols are present simply to ornament. The National Museum deploys symbols in a way that allows for recall. Hence the symbols are alive to their own possibilities. The possibilities in question, however, are linked to a complex sense of identity. Rather than symbols held by a unified singular idea (or its opposite, the absence of any idea, this opposition being the postmodern project), in this building symbols are dispersed in order to mark the presence of a necessarily dispersed subject. This is precisely what postmodernism in its use of the "dead" symbol could not achieve.

11. Jacques Derrida, "Question d'étranger: Venue de l'étranger," in Anne Dufourmantelle and Jacques Derrida, *De l'hospitalité* (Paris: Calman-Lévy, 1997), 29. Derrida's reference to the name, and the giving of names, is fundamental not just to the question of citizenship but to the way an introduction into a particular social setting took place. A freed slave in the Roman period was given a name. Naming therefore was not just a mode of welcoming, it was the sign of participation. See in this regard J. P. V. D. Balsdon, *Roman and Aliens* (London: Duckworth, 1979), 82–97.

12. Jacques Derrida, "Une hospitalité a l'infini," in *De l'hospitalité: Autour de Jacques Derrida,* ed. Mohammed Seffahi (Genouilleux: Éditions la Passé du Vent, 2001), 116.

13. An important recent historical work that pursues such a possibility is Mark McKenna, *Looking for Blackfella's Point: An Australian History of Place* (Sydney: University of New South Wales Press, 2002). What is of great interest in this study is the way in which what he describes as "acknowledging different histories" lends itself to a different set of cultural practices than would follow from attempts to base activity on unified national histories. This position is well worked out in the book's concluding section.

14. Rather than seek to define "cosmopolitan," its value resides in what it occasions. Once it names the interruption of the opposition national/international, then part of its force is that it necessitates a redefinition of both these terms. In the process of redefinition—a process that inscribes the actual place back into the activity of redefinition—there is now a politics of place, since the literal place is present in that politics works to determine policy, since place is not reducible to literal presence. In the opening that the literal stages, there is a site of activity. Occurring within it is a range of practices that, while related to national and international traditions, neither repeats them nor mimes their form. The cosmopolitan becomes therefore a site of activity, experimentation, and creation; a site defined by the maintaining and enacting of potential. This accounts for why even though this chapter makes continual references to the Australian situation, once place is taken in a cosmopolitan sense, then the Australia in question has to be refracted through a cosmopolitan lens. This chapter needs to be read as just such an attempt.

15. Stephen Muecke, "Outback," in *Imagining Australian Space: Cultural Studies and Spatial Inquiry,* ed. Ruth Barcan and Ian Buchanan (Perth: University of Western Australia Press, 1999), 128.

16. I have tried to develop this conception of the nation-state in a number of contexts. See in particular "Cosmopolitan Citizenship," *Kamena Supa* 1 (1995): 33–49; and "Modernity, National Identity, and the Diaspora," in *Alchemic Surrender,* ed. M. Kurza (Kiev: Soros Center for the Visual Arts, 1998), 18–22. The latter essay was reprinted in *Postcolonial Culture and Literatures: Modernity and the Commonwealth,* ed. Andrew Benjamin and Robbie Goh (New York: Peter Lang, 2001).

17. James Clifford, *Routes: Travel and Translation in the Late Twentieth Century* (Cambridge, Mass.: Harvard University Press, 1997), 252.

18. One of the most persuasive discussions of the way in which the state of affairs marked out in this chapter by the term "cosmopolitan" can be understood as that which interrupts the opposition between the national and the universal can be found in a paper by Jean-François Lyotard, "Histoire universelle et differences culturelles," *Critique,* no. 456 (May 1989). While most treatments of the term tend to conflate it with a version of internationalism, even its earliest uses involved greater complexity than such a move would evidence. In Plutarch's treatment of Zeno's *Republic,* the expression "all people" should be considered as being members of "one community and one polity." And yet it would be too easy to dismiss Plutarch's text (and by extension Stoicism) by virtue of what seems to be an unproblematic positing of universality. What needs to be undertaken is a reworking of this conception of universality within a framework constructed by the way in which "fortune" is displaced in favor of the philosophical. Philosophy here is neither Aristotelianism nor Platonism. It can be taken as the displacing of universality in the name of a conception of collectivity in which alterity plays a central role. For the text, see Plutarch, "On the Fortunes or the Virtue of Alexander," in the *Moralia,* trans. Frank Cole Babbit (Cambridge, Mass.: Harvard University Press, 1972). The passage in question is 6.329a–b.

Chapter 7

1. Clement Greenberg, "Manet in Philadelphia," in *The Collected Essays and Criticism,* ed. John O'Brian (Chicago: University of Chicago Press, 1993), 4:240. It should also be noted that Carol Armstrong begins her own excellent study of Manet with the same quotation. See her *Manet Manette* (New Haven, Conn.: Yale University Press, 2002), xi–xviii. My concern here is to offer another account of this "inconsistency."

2. Bataille's study of Manet was originally published in 1955 by Éditions d'Art Albert Skira. In this chapter the version in Bataille's *Oeuvres complètes* (abbreviated *OC*) will be cited. For an overview of his writings on art which, while concentrating on the Manet book, unfortunately does not connect that work to Bataille's enterprise as a whole, see Youssef Ishaghpour, *Aux origins de l'art moderne: Le Manet de Bataille* (Paris: Éditions de la Différence, 2002).

3. The term *sens* has recently been the subject of considerable philosophical reflection. (Special consideration should be given to the work of Jean-Luc Nancy. See in particular his *Le sens du monde* [Paris: Éditions de Galilée, 1993].) Part of the term's significance lies in its allusion as much to direction as to meaning. The term will continue to figure prominently in Bataille's attempts to identify limits and their overcoming. For Bataille the claim about Manet, once read within the framework of his other investigations of *sens,* introduces a critical perspective. Having a *sens* marked by apartness is to have a critical relation to a dominant tradition. This implicit affirmation of criticality should be assumed in the investigation of a conception of materialism that is not reducible to its metaphysical determination; i.e., caught in an oscillation between the ideal and its actualization (successful or not).

4. See in this regard Louis Marin, "Variations sur un portrait absent: Les autoportraits de Poussin: 1649–50," in *Sublime Poussin* (Paris: Éditions de Minuit, 1995), 186–209.

5. One of the most sustained presentations of this position is found in Stefan Germer, "L'ombre du peintre: Poussin vu par ses biographers," in *Les "vies" d'artistes,* ed. Matthias Waschek (Paris: Musée du Louvre, 1996), 105–24. Germer also describes the 1650 self-portrait as a "réflexion sur les possibilities de la representation" (109). The important point, however, is how this "réflexion" is understood. Victor I. Stoichita takes up the general question of shadows in painting, a question addressed in part by a study of Poussin's self-portraits, with great acuity. See his *A Short History of the Shadow* (London: Reaktion Books, 1997), 98–103. It should not be thought, however, that Poussin is unique in this regard. As is often remarked, the question of painting—of painting's presence to itself—also figures in aspects of Dutch painting in the seventeenth century. Willem van Haecht's *Studio of Apelles* and his *The Cabinet of Cornelius van der Gheest* would be central to any discussion of this point. (In regard to the second of these works, see Zirka Zaremba Filipczak, *Picturing Art in Antwerp 1550–1700* [Princeton, N.J.: Princeton University Press, 1987], 58–72.) The distinction, as Germer indicates, is between the incorporation of paintings within paintings, where it emerges as an end in itself, and as a "réflexion" on the act. (A reflection on painting presented as part of the painting.) The fact that Poussin stages this concern within a self-portrait, coupled to the solitary representation of Pittura, as opposed to a multitude of different paintings, when taken together delimit the particularity of the 1650 self-portrait.

6. Giovanni Pietro Bellori, *Le vite de pittori, scultori et architetti moderni* (Quaderni dell'instituto di storia dell'arte della universita di Genova, 1964).

7. Bellori, *Le vite de pittori,* 529.

8. The iconic presence of friendship in the hands on Pittura's head, as noted by Bellori, as well as in the ring on "Poussin's" hand, are discussed by Oskar Bätschmann in his *Nicolas Poussin: Dialectics of Painting* (London: Reaktion Books, 1996), 47–50.

9. I have discussed this aspect of Greenberg's work in my *What Is Abstraction?* (London: Academy Editions, 1996).

10. G. W. F. Hegel, *Vorlesungen über die Ästhetik* (Frankfurt am Main: Suhrkamp, 1986), 60.

11. There is a sense, therefore, that the problem of the "ready-made" and the *objet trouvée* is already present in Hegel's formulation. Such an object has to be at once reducible to its ordinary presence and transformed—allowed to become extraordinary—by the process that confers on it the quality of art.

12. Charles Baudelaire, "Salon de 1846," in *Oeuvres complètes* (Paris: Robert Laffont, 1986), 642.

13. The event of art is the movement—a movement inherent to the work of art—beyond the determinations given by authorship, chronologically determined historical contexts, envisaged and assumed audiences, etc. The event of art is captured as much by Bataille's formulation of Manet's work as having "a meaning [*un sens*] apart" as it is in loosening the hold of content and thus in releasing art's potential. An act—a release—occurring within and as criticism. The event of art is inevitably marked by, to use a term of Bataille's that will emerge at a later stage in this analysis, the "unforeseen" *(imprévu).* The unforeseen demands the attention of criticism. While it is not pursued in the same way, Derrida writes of the event in a similar vein by arguing that

> there is only an event where it is not awaited, there where it is no longer awaited, where the coming of that which happens interrupts the awaiting [*il n'y d'événment que là oú ça n'attend pas, oú on ne peut plus attendre, oú la venue de ce qui arrive interrompt l'attente*].

This quotation is from Jacques Derrida, "Une certaine possibilité impossible de dire l'événement," in Gad Soussana, Alexis Nouss, and Jacques Derrida, *Dire l'événement, est-ce possible?* (Paris: L'Harmattan, 2001), 84.

14. While it was Bataille who kept Benjamin's papers hidden, and thus safe, during the Second World War, and apart from some biographically driven studies, there has yet to be an analysis of the two men's respective political and aesthetic projects that seeks to plot points of connection and distance between them. An important beginning has been made by Martin Weingrad; see his "The College of Sociology and the Institute of Social Research," *New German Critique,* no. 84 (Fall 2001): 129–61. Despite drawing on a range of contemporary sources, Weingrad's analysis is still limited by the historical conjuncture rather than the potential of the two bodies of work.

15. I have tried to develop this interpretation of Benjamin in chapter 2.

16. Of the many possible examples, see Fritz Novotny, *Painting and Sculpture in Europe 1780–1880* (Harmondsworth, Eng.: Penguin, 1978). Novotny describes Manet's use of "motifs" from the history of art as occurring "in a manner which was irreconcilable with the current eclecticism and aroused violent passions" (332). Since little is to be gained by simply positing this setup, the question addressed throughout this chapter is how this irreconcilability is to be understood.

17. In this regard, see Nancy Locke, *Manet and the Family Romance* (Princeton, N.J.: Princeton University Press, 2001). What is important about her work is the way psychoanalysis is used to account for the relationship between the painter and the subjects within the paintings. While this is a perfectly justifiable approach, the difference here is that Bataille's interest in psychoanalysis lies in its capacity to provide an account of production. Materialism differs from empiricism in that the former is inevitably bound up with a process of production. It is the process that works to exclude the hold of the idea.

18. The overdetermined status of Raphael in these discussions needs to be noted. What is interesting with Bataille—and also with Manet—is that the relationship to Raphael is not structured by the positive and negative poles of imitation.

19. The whole question of Manet's relation to the history of painting has been well documented. The most important and most impressive account is Michael Fried's *Manet's Modernism* (Chicago: University of Chicago Press, 1996). Since the approach to Manet pursued here has been via Bataille and is centered on the question of appearance, Fried's overriding concern with situating Manet's work within his rewriting of the history of French art in terms of the opposition between theatricality and absorption has been left to one side. For a detailed engagement with Manet in terms of his contemporary critical reception—rather than his sources—see James H. Rubin, *Manet's Silence and the Poetics of Bouquets* (Cambridge, Mass.: Harvard University Press, 1994).

20. The most informative and influential study of Bataille's materialism is Denis Hollier, "The Dualist Materialism of Georges Bataille," trans. Hilari Alkred, *Yale French Studies,* no. 78 (1990): 124–39.

21. Goya remained an important part of Bataille's writing project. It should be noted, in this regard, that he planned a new project on Goya and Sade in 1960—less than two years before his death. See his letter to Dionys Mascolo in *Georges Bataille: Choix de Lettres: 1917–1962,* ed. Michel Surya (Paris: Gallimard, 1997), 542. For Bataille's sustained treatment of Goya, see *Oeuvres complètes,* 11:304–12, 550–55. A discussion of Goya also appears in Bataille's Manet book.

22. I have attempted such an approach to a philosophy of art in *Disclosing Spaces: On Painting* (Manchester: Clinamen Books, 2004).

23. The emphasis on process and thus on the economic was there in Freud's first formulation in the "Project for a Scientific Psychology," in *The Standard Edition of the Complete Psychological Works of Sigmund Freud,* ed. James Strachey (London: Hogarth, 1966), 1:295–346. This emphasis on processes is a consistent motif in Freud's thought. Its important reappearance is in section 5 of "The Unconscious" (*Standard Edition,* 14:186). The fundamental position is the formulation of the "core of the unconscious" as "consisting of representatives of the drives [*Triebrepräsentanzen*]," and that therefore the dynamic model is what accounts for presentation. Presentations are always the effect of the two psychical processes.

24. A point also made by Hubert Damisch in "Une femme, donc: 'Le déjeuner sur l'herbe,'" in his *Le jugement de Paris* (Paris: Flammarion, 1992).

25. Manet worked in a different direction. As Bataille notes, "Manet tournait le dos à la peinture de l'histoire" (*Oeuvres complètes,* 9:155).

26. The significant point here is that it will be a conception of realism that is defined beyond the ambit of representation. What Bataille is insisting on is another conception of realism. Most interpreters of realism in Manet still define its presence in terms of representational models. See, for example, Gregory Galligan, "The Self Pictured: Manet, the Mirror and the Occupation of Realist Painting" *Art Bulletin* 80, no. 1 (May 1998): 139–71.

27. The key text here is Zola's *Le roman expérimental* (Paris: Flammarion, 1969). One of the significant moments within that text is Zola's defense of Balzac's *Le cousin Pons* in terms of its being an experimental novel. The defense involved the argument that what Balzac had created was a setting in which experimentation—the investigation of recognizable character traits and worldviews—could take place. Experimentation and hence naturalism were inextricably bound up with the creation of settings, sites of experimental possibility. This is not to argue that Zola was not concerned with questions of innovation and therefore with problems of continuity and discontinuity. Rather, developments in literature took place by the literary adoption—perhaps adaptation—of the scientific methodology of Claude Bernard. In regard to the latter, see his *Introduction à l'étude de la médecine expérimentale* (Paris: Garnier-Flammarion, 1966). On this, see my "Experimentation as a Defence of Literature: Zola's *Le roman expérimental,*" in *The Cambridge Companion to Zola,* ed. Brian Nelson (Cambridge: Cambridge University Press, forthcoming).

28. Once the question of the subject is central, and this is the way caricature works by securing and demanding the subject's centrality, then there would be a need to distinguish between caricature and cartoons. For an important discussion of caricature and cartoons, though one which tends to conflate them rather than insist on their particularity, see E. H. Gombrich, "The Cartoonist's Armoury," in *Meditations on a Hobby Horse* (London: Phaidon, 1971), 120–40.

29. The relationship between the opening up of "sens" and an insistence on the informal has also been noted by Jean Clay, "Ointments, Makeup, Pollen," trans. John Shepley, *October* 27 (1983): 3–44. In a discussion of "speed" in Manet's work, Clay argues that

> speed is one of the agents that allows Manet to wrench the painting from the conditioning of a form that is already there. (31)

Whatever it is that this "wrenching" away stands for, it is linked to the absence of an already determined formal presence. It is this opening up that attests on the one hand to the structuring presence of the informal (discussed by Bataille as the "l'informe" [*Oeuvres complètes,* 1:217–18]) and on the other to the severance of style and appearance that marks aesthetic production in the modern period.

30. Bataille's most sustained discussion of poetry in these terms is found in his "L'éxperience intérieur" (see *Oeuvres complètes,* 5:156–58). Part of that discussion could be usefully linked to Blanchot's exploration of Mallarmé, especially the identification of the productive nature of negativity in the latter's poetic practice. See Maurice Blanchot, *L'espace littéraire* (Paris: Gallimard, 1955), 134–36.

31. I have pursued this aspect of Bataille's work, which could be described more generally as a conception of negativity that is productive, in "Figuring Self-Identity: Blanchot's Bataille," in *Other than Identity: The Subject, Politics and Aesthetics,* ed. J. Steyn (Manchester: Manchester University Press), 9–32.

32. Indeed, this move away from sacrifice as a determining logic and toward its reconceptualization in terms of the relationship between the informal and the finding of form allows for a shift in concern from a logic of sacrifice toward another politics. See Nancy, *Le sens du monde,* 142.

33. A similar interpretation of *The Fifer* is given by Rubin, who argues that the painting's title

> identifies not a figure from reality but a physical sign, for which the figure that was its referent now seems no more than a pretext.
>
> (Rubin, *Manet's Silence,* 65)

34. This is Bataille's translation of Hegel. The passage from the *Phenomenology of Spirit* from which it comes is the following. (It goes without saying that the nature of Bataille's engagement with Hegel warrants a study in its own right.) I have inserted the original German plus Bataille's translations where appropriate.

> Death . . . if we so wish to name that unreality—is the most terrible thing there is and to uphold the work of death is the task that demands the greatest strength. Impotent beauty [*Die kraftlose Schönheit, Le beauté impuissante*] hates this awareness, because understanding makes this demand of beauty, a requirement which beauty cannot fulfil. Now, the life of spirit is not that life which is frightened of death, and spares itself destruction, but that life which assumes death and lives with it. Spirit attains its truth only by finding itself in absolute dismemberment. [*Er gewinnt seine Wahrheit nur, er in der absoluten Zerrissenheit sich selbst findet. L'esprit n'obtien sa vérité qu'en se trouvant soi-meme dans le déchirement absolu.*] It is not that prodigious power by being positive that turns away from the negative, as when we say of something: this is nothing or (this is) false and, having (thus) disposed of it, pass from there to something else; no, Spirit is that power only to the extent that it contemplates the negative face to face and dwells with it *(bei ihm verweilt, séjourne pres de lui).* This prolonged sojourn [*Dieses Verweilen, Ce séjour-prolongé*] is the magical force which transposes [*umkehrt*] the negative into given-Being.

Index

Adorno, Theodor, 141n10, 150n2, 157n7
Akcan, Esra, 151n10
Allen, Stan, 152n13
Aragon, Louis, 59
Aristotle, 112, 157n8, 159n18
Armstrong, Carol, 159n1
Ashton Raggatt MacDougall, xxiii, 90–93, 96–97, 157n10

Balsdon, J. P. V. D., 158n11
Bataille, Georges, xviii, 63–65, 76, 124–38, 139n2, 141–42n11, 150n1, 159n2, 160–64nn17–34
Bätschmann, Oskar, 160n8
Baudelaire, Charles, xii, 128
Behrendt, Walter Curt, 39, 147n2
Behrens, Peter, 48, 150n18
Bellori, Giovanni Pietro, 125, 160n6
Benjamin, Andrew, 141n7, 143n15, 144n8, 147n12, 151n9, 159n16, 162n22, 163n27, 164n31
Benjamin, Walter, ix–x, xii–xviii, xxi, xxiii, 5–60, 111, 128, 139–40nn1–3, 141–42n11–13, 143–44nn1–4, 144–47nn1–12, 150n20, 161n14
  works: *Arcades Project, The,* ix–xi, xii, xvi, 6, 14, 16, 19–22, 23, 25–26, 32, 40, 50, 52, 54, 58–60, 143–44n3, 145n6, 146nn8–9
  Concept of Criticism in German Romanticism, The, 6, 7, 9, 11, 22–23
  "Experience and Poverty," 56
  "Goethe's Elective Affinities," 6–13, 14–15
  "On Some Motifs in Baudelaire," 14
  "On the Concept of History," xviii, 14, 16, 17–18, 20, 25–27, 29–38, 142n12, 145n5
  Origin of German Tragic Drama, The, 15–16, 146n8
  Response to Oscar A. H. Schmitz, 57–58
  "Two Poems by Friedrich Hölderlin," 6
  "Work of Art in the Age of Its Technological Reproducibility, The," 14, 17, 26
Bergdoll, Barry, 149n12
Bernard, Claude, 163n27
Bernik, Stane, 154n9
Bertall, Albert, 134
Bhabha, Homi, 112, 143n16
Blanchot, Maurice, 164nn30–31
Boetticher, Karl, 44–47, 78, 83, 84, 85, 149n14, 149n16
Boher, Karl Heinz, 139n3
Borromini, 148n9
Brecht, Bertolt, 26, 27
Breithaupt, Fritz, 145n4
Burkhardt, François, 154n10

Cacciari, Massimo, 157n5
Cadava, Eduardo, 145n4
Casey, Edward, 151n11
Castiglione, Baldassare, 42
Caygill, Howard, 143n2, 146–47n12
Clay, Jean, 163n29
Clifford, James, 120
Corngold, Stanley, 150n21
Corbusier, Le (Charles-Édouard Jeanneret), 92
Courbet, Gustave, 132
Curtis, William J. R., 154–55n14

Damisch, Hubert, 163n24
David, Jacques Louis, 132
de Chatelou, Paul Fréart, 124
de Meuron, Pierre, xxiii, 94–96, 154n14
Derrida, Jacques, 117–18, 158nn11–12, 161n13
Dolinar, Lojze, 80
Dyck, Robert Gilkey, 153n8

Ferris, David, 148n6
Filipczak, Zirka Zaremba, 160n5
Foster, Hal, 152n13
Frampton, Kenneth, xxii, 77–78, 143n17, 152nn13–15
Freud, Sigmund, 130, 162n23
Fried, Michael, 148n6, 162n19

Galligan, Gregory, 163n26
Germer, Stefan, 160n5
Goethe, Johann Wolfgang von, 6, 8, 9, 11, 17, 41, 146n8
Gombrich, E. H., 163n28
Gogh, Vincent van, 129
Goya, Francisco, 129, 162n21
Grabijan, D., 153n5
Greenberg, Clement, 124, 127, 160n9
Gundolf, Friedrich, 17

Haar, Michel, 151n12
Haecht, Willem van, 160n5
Hamacher, Werner, 145n4
Hansen, Beatrice, 143n1
Hazlitt, William, 148n7
Hegel, Georg Wilhelm Friedrich, 18, 19, 127–28, 137, 139n2, 140n4, 140–41n6, 161n10, 164n34
Heidegger, Martin, 77, 78–79, 151n12
Heraclitus, 155–56n4
Herzog, Jacques, xxiii, 94–96, 154n14
Hittorff, Jacques Ignace, 39, 147n4
Hölderlin, Friedrich, xiv, 6, 8, 10, 38, 141n8
Hollier, Denis, 150n1, 162n20
Horkheimer, Max, 157n7
Hübsch, Heinrich, xvii, 39, 31–42, 44, 46–48, 51, 52, 58, 82–83, 140n5, 147n1, 148n7, 149n15

Ibsen, Henrik, 52, 53
Ishaghpour, Youssef, 159n2

Jacobs, Carol, 144n4
Jager, Janez, 154n9
Jugendstil, xii, xvi, xvii, 34, 39–40, 48–57, 78
Jünger, Ernst, 142n13

Kant, Immanuel, xv, 18, 23–24, 144n9, 152n1
Kipnis, Jeffrey, 155n14
Kraus, Karl, 37
Krecic, Peter, 153n4
Krell, David Farrell, 145–46n7

LAB Architectural Studio, xxiii, 97–105
Laplanche, Jean, 151n9
Locke, Nancy, 162n17
Löhneysen, Wolfgang von, 148n8
Lokan, Andrew, 155n3
Longinus, 42, 148n10
Loos, Adolf , 66, 151n5
Lotze, Hermann, 60
Löwith, Karl, 145–46n7
Lyons, 94, 154n13
Lyotard, Jean-François, 154n11, 159n18

Macarthur, John, 154n12
Mackay, Angus, 157n5
McKenna, Mark, 158n13
Mallgrave, Harry Francis, 149n17

Malpas, Jeffrey, 151n11, 155n1
Malraux, André, 136
Manet, Édouard, xiv, xv, xviii, 124–38, 159–60nn1–3, 161n13, 162nn17–19, 162n21, 163nn25–26, 163n29
Marin, Louis, 160n4
Marx, Karl, 29, 35, 60, 152n2, 156n5
Mascolo, Dionys, 162n21
Michelangelo, 42, 87, 148n9
Muecke, Stephen, 119–20

Nancy, Jean-Luc, 160n3, 164n32
Nietzsche, Friedrich Wilhelm, xviii, xix, 68–76, 79, 141–42n11, 145–46n7, 147n4
Novotny, Fritz, 161n16

Osborne, Peter, 141n10
Ostwald, Michael J., 154n13

Patton, Paul, 155n3
Pirenne, Henri, 152n2, 156n5
Plato, 42, 159n18
Plečnik, Jože, xviii–xix, xxiii, 84–89, 90–91, 105, 153–54nn4–10
Plutarch, 159n18
Pointel, John, 125
Poussin, Nicolas, 124–26, 160n5, 160n8
Prelovsek, Damjan, 153n4
Proust, Marcel, 139n3

Quintilian, 126

Raphael, 42, 43, 129, 148n8, 149n11, 162n18
Ravnika, Edo, 153n7
Reynolds, Joshua, 148n7
Riegel, Alois, 153n5
Robespierre, Maximilien, 29, 35
Rosenthal, Carl Albert, 149n15
Rubin, James H., 162n19, 164n33
Ruff, Thomas, 94
Russell, D. A., 148n10

Sade, Marquis de, 162n21
Schelling, Friedrich Wilhelm Joseph von, 148–49n11
Schinkel, Karl Friedrich, 44–45, 56
Schlegel, Friedrich von, 9, 23, 144n8
Schmitz, Oscar A. H., 57
Schwarzer, Mitchell, 149n14
Semper, Gottfried, 85, 94, 149n17
Sereni, Emilio, 157n5
Smart, Bates, 97
Stoichita, Victor I., 160n5

Tiedemann, Rolf, 146n12
Titian, 129

Valéry, Paul, 132–33
Velde, Henry, van de, 49–50, 56, 150n19
Vurnik, Ivan, 154n9

Wagner, Otto, 84, 94
Ward, Janet, 146n10
Weingrad, Martin, 161n14
Whittaker, C. R., 157n5
Wigely, Mark, 147n4
Winckelmann, Johann Joachim, xvii, 41–44, 51, 52, 147–48nn6–7, 148–49n11
Wollen, Peter, 144–45n1

Zanten, David Van, 147n5
Zeno, 159n18
Zola, Emile, 132–33, 163n27

About the Author

Andrew Benjamin is a professor of critical theory in design and architecture at the University of Technology Sydney (Australia). He has also taught philosophy and architectural theory in both the United States and Europe, where he teaches regularly at the Architectural Association in London. His recent books include *Architectural Philosophy: Repetition, Function, Alterity; Philosophy's Literature;* and *Disclosing Spaces: On Painting.* He is joint general editor of the Walter Benjamin Studies Series.

Avant-Garde & Modernism Studies